Information Technology and Global Governance

Series Editor
Derrick L. Cogburn
American University
Bethesda, MD, USA

Information Technology and Global Governance focuses on the complex interrelationships between the social, political, and economic processes of global governance that occur at national, regional, and international levels. These processes are influenced by the rapid and ongoing developments in information and communication technologies. At the same time, they affect numerous areas, create new opportunities and mechanisms for participation in global governance processes, and influence how governance is studied. Books in this series examine these relationships and influences.

Ignas Kalpokas

Technological Governance and Escapism in Times of Accelerated Change

palgrave
macmillan

Ignas Kalpokas
Public Communication
Vytautas Magnus University
Kaunas, Lithuania

ISSN 2946-3297 ISSN 2946-3300 (electronic)
Information Technology and Global Governance
ISBN 978-3-031-60889-6 ISBN 978-3-031-60890-2 (eBook)
https://doi.org/10.1007/978-3-031-60890-2

This Palgrave Macmillan imprint is published by the registered company Springer Nature Switzerland AG.
The registered company address is: Gewerbestrasse 11, 6330 Cham, Switzerland

Paper in this product is recyclable.

This book is dedicated to my son Samuelis Konstantinas, who was born early enough to witness (although not necessarily appreciate) much of this book's genesis.

About the Book

This book examines escapist coping strategies brought about by the pace, breadth and governance of technological change. It argues that escapism manifests in various forms, ranging from nostalgia for a fantasised past of unhindered reason and agency, to progressive visions of societal and political improvement and greater empowerment. Drawing on post-humanist theory and critical disability studies, the book also assesses how escapism should not be viewed as an unavoidable reaction to technological change, and develops a model for an ethically equitable relationship between humans and technology. It will appeal to all those interested in media and communication studies, technology studies, philosophy, governance and politics.

Contents

About the Author

Ignas Kalpokas is Associate Professor in the Department of Public Communication, Vytautas Magnus University, Lithuania. He also holds research positions at LCC International University (Lithuania) and SGH Warsaw School of Economics (Poland). His research focuses on the social and political impact of digital technologies, fake news and information warfare, and media theory.

CHAPTER 1

Introduction

Abstract While, at first sight, it might transpire that the prevalent views on technological change and, correspondingly, predictions for the future are diverse, upon closer inspection, they tend to coalesce into two broad camps. On the one side, views that can be generally described as technological progressivism provide an optimistic—even utopian—vision of technology solving humanity's problems and enabling full mastery over the world. On the other side, though, technological pessimism engages in nostalgia and paints the future in dystopian terms, mourning the alleged loss of human autonomy and centrality vis-à-vis the world. Both views are interpreted here as coping strategies in the face of rapid technological change that are more illustrative of human propensities than of technology itself.

Keywords Coping strategies • Technological change • Progressivism • Nostalgia

Rapid technological change has been a key characteristic of contemporary societies, especially with regard to digital technologies. In particular, though, the nexus of datafication, platformisation, and artificial

I. Kalpokas, *Technological Governance and Escapism in Times of Accelerated Change*, Information Technology and Global Governance, https://doi.org/10.1007/978-3-031-60890-2_1

intelligence (AI) has been more prominent than other changes—indeed, undergirding the others. It is, therefore, unsurprising that such technological transformations are accompanied by strong beliefs and increasingly entrenched discursive practices pertaining to their effects and, particularly, to their impact on the individual and societal levels (see e.g. Hoff, 2023). However, the framing of these technological developments has been far from universal, albeit in a peculiar way: while, at first glance, it seems like diverse perspectives abound, upon closer inspection, they tend to align into two broad groups. The first of such groups could be called nostalgic: such views, in their own internally diverse ways, posit technological change as largely dangerous and bringing about harmful side effects (or, in more radical variations, bringing harm as the *main* effect). Such theories tend to lament the pronounced erosion of the rational autonomous human agent, with both the present and the future broadly understood as times of agentic, cognitive (and sometimes even moral) degradation. Meanwhile, the second group of frames can be labelled techno-progressivist. Here, an optimistic, over outright utopian, view of technologically enhanced future dominates. Instead of perceiving technological change as bringing about human degradation, these theories often tend to foresee augmentation of the human as the most likely state of affairs, thereby increasing natural capacities.

Here it is fruitful to keep in mind the literature on sociotechnical imaginaries. Notably, Jasanoff and Kim (2009) and Jasanoff (2015) describe such imaginaries as collectively imagined visions of social life, particularly oriented towards the future, that are held, stabilised performed within dominant social and institutional apparatuses, and pertain to technology as the presumed vehicle of fundamental transformations. Similarly, Lupton (2020: 15) defines such imaginaries as 'constellations of speculation and narratives that draw on understandings, norms, values and experiences of the world' while Sörum and Fuentes (2023: 25) see them as 'cultural resources for making sense and guiding people's interpretation of a practice, an experience or a new technology'. At first glance, Marčetić and Nolin (2023: 94–95) may even seem to have cut the work of this book short by providing definitions for both utopian and dystopian sociotechnical imaginaries that largely correspond with the progressivist and nostalgic tendencies identified in this book: while utopian imaginaries for them depict emerging technologies 'as capable of improving social systems and economies', dystopian ones are focused on scrutiny and critique.

Nevertheless, the preceding account is seen here as relatively mild and failing to encompass the existential anxieties (seen as either emergent or soon-to-be overcome), inherent in much of the contemporary debates about new and emerging technologies. Such anxieties are subsequently traced in this book to an attempt at the preservation of the anthropocentric worldview that, while both empirically and normatively wrong, has held a tight grip over Western thought.

It is indeed possible to claim that technological innovations have a 'seductive' character that drives towards either utopian or dystopian extremes (see e.g. Preece et al., 2022). Notably, as Pink et al. (2022: 1) emphasise, discussions of new technologies tend to be 'shrouded with narratives which highlight extreme and spectacular examples, rather than the ordinary mundane realities that characterise the overwhelming majority of people's actual encounters with them'. Arguably, it is this thrust towards the spectacularisation of technological change—either things going spectacularly wrong or, to the contrary, displaying (usually the promise of) dramatic improvement—that renders dichotomic narratives seemingly more intuitive. Nevertheless, being intuitive is not the same as being productive and ethically justifiable, thereby only further underscoring the need for an approach not based on extremes but also one that does not succumbs to a mere synthesis.

The tendency of technology frames, or imaginaries, to coalesce into such broad and dichotomous groups is explained in this book as a coping strategy. In particular, it is claimed that technological change undermines the anthropocentric way of thinking, characteristic of the Western tradition. This undermining takes place not because such anthropocentrism ever was ontologically or ethically justifiable but *precisely* because it never was. Effectively, technology acts as a mirror that reflects back the absent centre of 'the human', resulting in a traumatic experience. It is here that the coping strategies kick in: the impossible anthropocentric model is either nostalgically looked back to or rescheduled for the future.

As for coping strategies in general, the very presence of such reactions is clearly not new and exclusive to the current time of rapid change. Instead, coping can be defined as 'a behavioural reaction to aversive situations, i.e. situations that induce psychological stress reactions' (Wechsler, 1995: 124). Similarly, according to Beaudry and Pinsonneault (2005: 494), coping comprises of 'the adaptational acts that an individual performs in response to disruptive events that occur in his/her environment'. Hence, seeking for coping methods and strategies can be understood as hardwired through evolution while lack thereof would reduce animal fitness and, therefore, chances of survival (Wechsler, 1995). Nevertheless, the particular strategies used in a given situation will not be uniform (let alone strategies adopted in different situations). Instead, the adoption of particular coping strategies is determined on an individual level, depending on the context of their engagement with, most recently, technology (Beaudry & Pinsonneault, 2005). Crucially, fundamental technological change is typically perceived as disruptive and, therefore, threatening or, at least, challenging, thus implying the need to cope (see, generally, Li et al., 2017), even if the coping mechanisms adopted may not be taken for granted as necessarily addressing the root causes of the problems involved (Benamati & Lederer, 2001: 199). In effect, then, establishing coping mechanisms sometimes might be (subjectively) more important than the efficiency of coping itself. Notably, coping behaviours as such tend to rest on definitions of the emerging situation through dichotomies between good and bad, virtuous and malicious (see e.g. Liang & Xue, 2009). Hence, a general trend that emerges throughout history whenever fundamental technological transformations are involved is to treat them either as an unconditional good (providing efficiency, control, freedom, wealth etc.) or as an unconditional bad (reducing human capacities and agency and potentially causing civilisation-wide harm); as such, reactions to digital technologies are also by no means new (Mick & Fournier, 1998).

Coping strategies have a tendency to follow the hedonic principle of seeking pleasure and trying to avoid pain—including in attempts to balance the positive and negative externalities of digital technologies (Liang & Xue, 2009). Nevertheless, as shown in this book, pleasure and pain can be placed at exactly the opposite ends of competing coping strategies. What the hedonic principle implies, nevertheless, is an escapist trend in order to 'handle stress and strains of daily life' (Han et al., 2022: 2). A crucial tenet here is 'a need of people wanting to "leave" the real world in which they live, both in a cognitive as well as emotional way' (Han et al., 2022: 3). Again, it will be seen that the nostalgic and progressivist escapisms have their 'exists' in opposite directions, in terms of both values and chronology. Nevertheless, the escapist outlook itself is shared.

It must be admitted that other authors have also focussed on dichotomic thinking about technological change, albeit in different ways. For example, Marušič and Sergaš (2023) also propose a two-pronged vision of the future, this time based on the figures of the basilisk and the zombie. Their vision, nevertheless, focuses on a double dystopia: enslavement of humans through the development of superior AI (basilisk) or cognitive automation of the entire humanity (turning humans into zombies). Hence, what they present is a double vision for a nostalgic escapism of a lost (or soon-to-be lost) world of (supposed) human mastery and autonomy. Meanwhile, Brynjolfsson and Unger (2023) manifest the opposite trend: while also positing two potential paths, and even identifying a positive and negative outcome, they nevertheless write from a standpoint of progressivist escapism. For them, the positive outcome (greater productivity and growth) is the default one *unless* technological progress is mismanaged, which would then bring greater inequality. In a similar way, AbuMusab (2023) posits technological development to be one of progressive improvement which, depending on regulatory choices, could end up as either a post-work utopia characterised by abundance or human replacement and increase in poverty. However, this book is different in the sense that it provides a comprehensive view of *both* the nostalgic and progressivist strategies and not different pathways within either of them.

Meanwhile, Pink et al. (2022: 3–4) come very close to the framework of this book by way of emphasising the optimist-pessimist dichotomy at the heart of current thinking about technology. Nevertheless, where they fall short is in terms of seeking to reestablish the anthropocentric framework on which that same dichotomy rests: namely, their call to 'rehumanise' technology very strongly rests on anthropocentric assumptions that are criticised in this book. In fact, should one choose to go back in time rather significantly, observations that dichotomic thinking is characteristic of attempts at conceptualising technology can be traced back all the way to Ada Lovelace, writing in 1843. According to her, '[i]n considering any new subject, there is frequently a tendency, first, to overrate what we find to be already interesting and remarkable and, secondly, [..] to undervalue the true state of the case, when we do discover that our notions have surpassed those that were really tenable' (Lovelace, c.f. Natale & Henrickson, 2024: 1914). And while this description better defines what we now know as the hype cycle, the propensity towards extremes is still clear.

The outlook proposed in this book is neither a synthesis of existing progressivist and nostalgic points of view (particularly as it is understood in dialectics) nor the result of some process of deliberation. In fact, particularly the latter option might end up not only as a sloppy escape but, even more importantly, as a mere reiteration of the anthropocentric assumptions inherent within the existing dichotomy. For example, Fiesler (2021: 18) provides a clear illustration of the preceding by resorting to collective foreseeing of the potential impact of emerging technologies, with the expectation that in the end, individuals will be able and willing to 'listen to each other, in order to create collective visions of the future' so that the voices of a multiplicity of diverse stakeholders are included. While, at first glance, the invocation to '[i]magine what the world might look like if everyone who touched technology examined it critically, in a creative, forward-thinking way' (Fiesler, 2021: 18) might seem highly attractive and inclusive, in fact, the latter quality is doubtful. Instead, it pitches a highly idealised socio-technological imaginary that occludes power, knowledge, and access asymmetries; posits a universal luxury to care and, in a typical fashion, re-renters human reason as a central ordering force in the world.

Of course, further alternatives are also present, including attempts to downsize the scale of the transformations themselves. For example, the term 'Artificial Intelligence' is itself not without controversy, with critics pointing out that there is nothing properly intelligent about AI: in fact, it is, according to this view, best understood as a simply more elaborate version of previous digital tools, without any ambit for independent action of its own (see e.g. Lanier, 2023). For others, much of the current debate is driven by 'algorithmic uncanny' (Balanzategui, 2023)– over-ascription of features and capacities to algorithms and other digital tools, especially by way of framing them as not only moderators of daily lives but also core shapers of the conditions for daily lives. Nevertheless, such views remain a minority. Neither are they offered as a way out of the dichotomy of nostalgia and pessimism. Instead, a model of interactive co-constitution between humans and digital artefacts, loosely based on posthumanist approaches, is suggested in this book.

This book is structured as follows. First, the need for coping strategies is set out by focusing on how digital technologies challenge some of the cherished assumptions about what constitutes the human. Then, focus shifts one of the most clearly evident examples of nostalgia—theories of post-truth. While work on post-truth does not have to be nostalgic as such, the prevalent view definitely reflects this coping strategy. The following chapter focuses on divergent interpretations of the impact of emerging technologies. Subsequently, projects for the metaverse are analysed as the ultimate expression of progressivist views—a space for unshackling the technologised human. Finally, the book ends with a search for an alternative non-binary approach that would resist giving in to either of the two extremes.

References

AbuMusab, S. (2023). Generative AI and Human Labor: Who is Replaceable? *AI & Society*. https://doi.org/10.1007/s00146-023-01773-3

Balanzategui, J. (2023). 'Disturbing' Children's YouTube Genres and the Algorithmic Uncanny. *New Media & Society*, 3521–3542.

Beaudry, A., & Pinsonneault, A. (2005). Understanding User Responses to Information Technology: A Coping Model of User Adaptation. *Management Information Systems Quarterly, 29*(3), 493–524.

Benamati, J., & Lederer, A. L. (2001). Rapid Information Technology Change, Coping Mechanisms, and the Emerging Technologies Group. *Journal of Management Information Systems, 17*(4), 183–202.

Brynjolfsson, E., & Unger, G. (2023, December). The Macroeconomics of Artificial Intelligence. *Finance & Development*. https://www.imf.org/en/Publications/fandd/issues/2023/12/Macroeconomics-of-artificial-intelligence-Brynjolfsson-Unger

Fiesler, C. (2021). Innovating Like an Optimist, Preparing Like Pessimist: Ethical Speculation and the Legal Imagination. *Colorado Technology Law Journal, 19*(1), 1–18.

Han, D.-I. D., Bergs, Y., & Moorhouse, N. (2022). Virtual Reality Consumer Experience Escapes: Preparing for the Metaverse. *Virtual Reality, 26*, 1443–1458.

Hoff, J.-L. (2023). Unavoidable Futures? How Governments Articulate Sociotechnical Imaginaries of AI and Healthcare Services. *Futures, 148*, 1–13.

Jasanoff, S. (2015). Future Imperfect: Science, Technology, and the Imaginations of Modernity. In S. Jasanoff & S. H. Kim (Eds.), *Dreamscapes of Modernity: Sociotechnical Imaginaries and the Fabrication of Power* (pp. 1–33). University of Chicago Press.

Jasanoff, S., & Kim, S. H. (2009). Containing the Atom: Sociotechnical Imaginaries and Nuclear Power in the United States and South Korea. *Minerva, 47*, 119–146.

Lanier, J. (2023, April 20). There Is no AI. *The New Yorker*. https://www.newyorker.com/science/annals-of-artificial-intelligence/there-is-no-ai

Li, G., Hou, Y., & Wu, A. (2017). Fourth Industrial Revolution: Technological Drivers, Impacts and Coping Methods. *Chinese Geographical Science, 27*(4), 626–637.

Liang, H., & Xue, Y. (2009). Avoidance of Information Technology Threats: A Theoretical Perspective. *Management Information Systems Quarterly, 33*(1), 71–90.

Lupton, D. (2020). 'Not the Real Me': Social Imaginaries of Personal Data Profiling. *Cultural Sociology, 15*(1), 3–21.

Marčetić, H., & Nolin, J. (2023). Utopian and Dystopian Sociotechnical Imaginaries of Big Data. *Journal of Digital Social Research, 5*(4), 93–125.

Marušič, J. Ž., & Sergaš, U. (2023). The Basilisk and the Zombie: Exploring the Future of Life with AI Through the Medium of Popular Culture. *Darbai ir Dienos/Deeds and Days, 79*, 95–113.

Mick, D. G., & Fournier, S. (1998). Paradoxes of Technology: Consumer Cognizance, Emotions, and Coping Strategies. *Journal of Consumer Research, 25*, 123–143.

Natale, S., & Henrickson, L. (2024). The Lovelace Effect: Perceptions of Creativity in Machines. *New Media & Society, 26*(4), 1909–1926.

Pink, S., et al. (2022). Everyday Automation: Setting a Research Agenda. In S. Pink et al. (Eds.), *Everyday Automation: Experiencing and Anticipating Emerging Technologies* (pp. 1–19). Routledge.

Preece, C., Whittaker, L., & Janes, S. (2022). Choose Your own Future: The Sociotechnical Imaginaries of Virtual Reality. *Journal of Marketing Management, 38*(15–16), 1777–1795.

Sörum, N., & Fuentes, C. (2023). How Sociotechnical Imaginaries Shape Consumers' Experiences of and Responses to Commercial Data Collection Practices. *Consumption Markets & Culture, 26*(1), 24–46.

Wechsler, B. (1995). Coping and Coping Strategies: A Behavioural View. *Applied Animan Behaviour Science, 43*, 123–134.

CHAPTER 2

Artificial Intelligence and Going Beyond Cherished Assumptions

Abstract As artificial intelligence (AI) and related technologies develop, they increasingly question perceived human primacy and autonomy. This is particularly notable given that anthropocentric assumptions lie at the core of the Western intellectual tradition, portraying humans as uniquely active and their environment as submissive and open to exploitation. Such ideas have reached the present courtesy of the Enlightenment, with Cartesian dualist thought being paradigmatic. Hence, this chapter engages with posthumanist critique, which seeks to replace hierarchic power relations with an egalitarian horizontal alternative characterised by interdependence. It thus becomes evident that cherished assumptions about human exceptionality and independence are being challenged by AI, giving rise to the need for coping strategies. Nevertheless, as the posthumanist critique demonstrates, such lack of exceptionality has always been the case.

Keywords Anthropocentrism • Enlightenment • Posthumanism • Agency • Interdependence

While questions of 'human nature' and supposedly 'essential' human features are perhaps bound to remain controversial, at least in the Western

I. Kalpokas, *Technological Governance and Escapism in Times of Accelerated Change*, Information Technology and Global Governance, https://doi.org/10.1007/978-3-031-60890-2_2

tradition, some taken for granted assumptions, usually concerning a unique human place in the world, can be seen to dominate. However, such assumptions are challenged by an ever-growing role played by AI and related technologies. By demonstrating capacities previously seen as uniquely human or challenging qualities such as autonomy and self-determination, AI destabilised and puts into question dominant images of the human. In order to better understand the nature of such assumptions and the discomfort (if not trauma) caused by digital technologies, it is useful to delve into posthumanist critiques of anthropocentrism and the exclusionary exploitative ways of thinking inherent therein. It turns out, therefore, that what causes the coping reaction is not technology directly, or at least not technology in isolation, but the mirror image a lack of anthropocentric fulfilment.

2.1 What Has AI Ever Done for Us?

Broadly speaking, one could define AI as 'the ability of a non-natural entity to make choices by an evaluative process' (Turner, 2019: 16). While the preceding definition transpires to be rather value-neutral, as the 'intelligence' bit suggests, the overall aim is typically the simulation of qualities that are typically assumed to be human and doing so in a way that exceeds human capacities in specific narrowly defined domains. After all, if mere replication of human performance was at stake, there would be no incentive to develop AI tools as no savings and efficiencies would be made. As a corollary, then, the driving assumption behind the adoption of AI lies in a seemingly simple and straightforward promise that the newly built tools would be capable of performing the same tasks 'more accurately, more cheaply and more quickly than humans do them today', with the latter leading directly to the possibility of substitution whereby '[o]nce AI can deliver on this promise, there will be no practical economic justification for humans to continue to be involved in many fields'—in fact, if one follows this logic, 'it would be affirmatively harmful and wasteful to have a human in the loop' (Toews, 2021). It is precisely in the replacement of humans with AI that the real economic value is seen to lie (Toews, 2021). While there are good reasons to criticise the economic reductionism of this view, it is, nevertheless, representative of a broader sentiment, prevalent in the technology sector and parts of the policy community.

In important ways, AI-enabled technologies challenge the assumed simplicity of the human/non-human, agent/subject distinction (Denhert

& Gunkel, 2023). At the very least, this involves *projecting* agency to AI and behaving towards it as if it had the relevant capacities (see e.g. Magee et al., 2023). Simultaneously, digital technologies, be it AI or machine learning more broadly, are often presented through the lens of agency combined with impermeability, thus framing them as decentring the human subject (see e.g. Fazi, 2020). Notably, digital agents occupy a paradoxical position: simultaneously possessing threatening potency (as in the discourse on substitution) and quasi-submissive lack of self-control (as in the discourse on inadequacy). In both cases, the net result is a continuation of the 'othering' and boundary-maintenance practices by way of framing digital technologies, and AI in particular, in relation to human agents and deciding the value of the former accordingly. Likewise, the performance of AI is typically framed with regard to what could be called 'cherished assumptions', that is, seemingly taken-for-granted assumptions about what does and does not count as human. Once such assumed boundaries are crossed, the net result transpires to be 'dehumanisation' through AI and robotics, which is typically seen in a negative light (see e.g. Zawieska, 2020), even though, as argued below, such boundary crossing merely demonstrates the artificiality of the boundaries themselves.

Problems of value and decision alignment are also crucial, with the goal effectively always boiling down to how AI can be best aligned with *human* values (or, rather, a particular interpretation thereof). It is notable that value alignment tends to be seen vis-à-vis some *version* of what is taken to represent universal human values, with the latter typically referring to the same independent and rational human ideal to which technology must serve (see e.g. Gabriel, 2020). Similar suggestions can be found with regard to proposals for 'human-centric AI': namely, subjection of the superior capacities of AI to assisting (or, perhaps, giving rise to) human flourishing (see e.g. Horvatić & Lipić, 2021). The element of service and enhancement—enjoyment of AI-assisted or AI-delegated decisions in line with the individual's true preferences or 'unprecedented opportunities for self-improvement'—clearly constitutes a significant proportion of the perceived attractiveness of deeper engagement with AI (see e.g. Puntoni et al., 2021: 132). In this sense, the potentially threatening capacity of AI to outperform humans can be reframed in a positive light as the enabler of progress and improvement of human life (Peeters et al., 2021). Likewise, enhancement is not necessarily seen as competitive—instead, from a human-centric perspective, there is simply too much of a difference between human and artificial intelligence and, therefore, substitution is, strictly speaking, impossible; what AI is thus seen as capable of doing is assistance to humans as well as freeing up time by performing less-fulfilling tasks (Peeters et al., 2021: 221). On such occasions, humans are typically imagined to still be either in the loop (exerting strict but benevolent control) or on the loop (setting overall human-focused goals) and, therefore, retaining their nature as the ultimate benchmarks (Mellamphy, 2021). Nevertheless, one element remains the same: the positioning of AI vis-à-vis human values and benefit.

In important ways, the human-as-a-benchmark narrative has been deeply engrained, particularly in European intellectual history and other traditions derived from it. Notably, beginning at least with Ancient Greece, and particularly since the dawn of humanism during the Renaissance and Enlightenment periods, 'ethical theories tend to be grounded in anthropocentric terms that privilege the achievement of human ends by way of human rationality at the expense of nonhuman lives' (Mellamphy, 2021: 13). There are clear tendencies in the human-AI relationship to stress conventional hierarchies: for example, when framing the relationship between humans and smart assistants, 'an unquestioned social hierarchisation, which justifies the unequal treatment of the machine' is typically established (Lind & Dickel, 2023: 2). Typical qualities thereby emphasised include submissiveness, following commands, and carrying out assistive roles, with the human being the one who 'requests, demands, and orders' (Lind & Dickel, 2023: 12–13). Notably, however, this relationship tends to become more egalitarian when such assistants are anthropomorphised (Lind & Dickel, 2023: 2), which also signals to the way in which the very idea of legitimate communication partners is anthropomorphised. Crucially, though, as intelligent assistants and AI-powered chatbots acquire increasing levels of interactive capacities, being able to act as companions, trusted advisers, friends, and, in some cases, even quasi-romantic partners, the relationships thus established have the power to dissuade users from interactions with fellow humans, particularly unfamiliar ones, thereby potentially leading to some level of substitution (Xia et al., 2024; see also Denhert & Gunkel, 2023). Likewise, as Ramati (2024) demonstrates, the addition of natural language processing (NLP) capacities adds additional perceived communicative agency (with the perception of the latter increasing proportionately to the human-likeness of the AI assistant's output), thereby at least superficially challenging the assumed communicative exclusivity of humans. Nevertheless, such substitution could also include aspects of playing out overtly sexualised (Jensen, 2023) or violent and otherwise toxic behaviours (Taylor, 2023). Hence, the general trend would involve emphasis on human superiority when then artificiality of AI chatbots and companions are visible, leading to substitution when a threshold of similarity is passed, but only through mirroring of either insecurities or toxic attitudes back to the humans.

However, digital technologies are also often seen as deeply implicated in challenging human autonomy and decision-making capacity or even processes of self-formation. One notable way of doing so is by digital tools performing mediation of experience and memory through platform affordances and enforcing identity management through logins and interfaces (Gardner & Tanenbaum, 2023). In this way, digital artefacts become if not fully fledged co-creators of identity then at least shapers of conditions for the latter's development and performance. On an even more fundamental level, digital technologies are increasingly challenging the idea of human autonomy and independence through augmentations, such as brain implants, currently mostly for medical treatment purposes, but also going beyond such use in the future (see e.g. Hamzelou, 2023). In this way, digital technologies become part and parcel of both the embodied and mental attributes of the human.

Even data which—despite (as shown later in this book) occupying a controversial place in the human-technology relationship—can, in many ways, be posited as an authentic link between the human person and the digital ecosystem is put into question in terms of its unique mode of representing a given individual; that is especially the case at times when existing data can be supplemented with synthetic data (see e.g. Jacobsen, 2024), thus implying a growing technological impossibility for humans to be even data-identical to themselves and to have a shaping power over the world through data. Datafication processes also find their way into domains previously considered quintessentially human, such as artistic creativity. In this case, human-created artworks are turned into data representations of themselves for the purpose of AI model training. Consequently the exclusivity of humans as creators of art and other cultural artefacts is increasingly challenged by generative AI, opening up previously seemingly taken for granted assumptions about authorship, creativity, and art-ness (see e.g. Kalpokiene & Kalpokas, 2023; Friedman, 2024). In the same vein, just like humans become alienated from their data by way of synthetic data, they are also alienated from the fruits of their own artistic labour as they are re-rendered for AI purposes.

To some extent, one could even witness AI as challenging established assumptions about life and death, including claims of being able to, in a way, bring the dead back to life via computer-generated animations and similar means (Kopelman & Frosh, 2023). In this way, AI becomes capable of putting the linearity of time into question: creating an 'algorithmic *as if*', not least by predicting and automating desires not only for alternative futures but also for alternative pasts (Kopelman & Frosh, 2023). Hence, even some ultimate point of reference that seemingly structure human life are rendered questionable.

2.2 Where Next? Excursus on Artificial General Intelligence and Superintelligence

Moving forward, one must also take account of the discussions concerning Artificial General Intelligence (AGI) and superintelligence. Here one immediately encounters a paradox: while AGI is often discussed, a clear definition of it is lacking (see e.g. Mitchell, 2024). A general understanding, though, is that while current AI (at the time of writing) is highly specialised and can exceed humans in specific tasks but cannot be extrapolated across the spectrum of cognition without being re-trained from scratch, AGI would perform across any conceivable tasks in a way that rivals and exceeds human capacity (see e.g. Baum & Villasenor, 2023; Salmon et al., 2023). Of course, the very classificatory scheme here reveals clearly pronounced anthropocentric tendencies: the human is, very literally, the measure of all things. The expectation is also that AGI would become a stepping stone towards an intelligence explosion whereby it would be able to create ever more intelligent versions of itself and ultimately reach what is known as superintelligence (see e.g. Graham, 2022). Indeed, the cherished assumption of human superiority is clearly at threat here.

Unsurprisingly, then, AGI and superintelligence can be defined in terms of both coping strategies discussed in this book, that is, progressivist and nostalgic tendencies. On the one hand, the debate around AGI can be seen as a progressivist embodiment of the overarching focus on human inventive and creative capacities—an almost God-like creation of advanced intelligence out of nothing. All of this is the case even in light of evidence that an act of technological inventorship might not be enough as experiential, social cultural factors are increasingly acknowledged as conditions sine qua non of human-like or superior intelligence (see e.g. Mitchell, 2024). To a large extent, it is popular to imagine that AGI would effectively be the last invention humans would have to make as it would then take on the task of the ultimate problem solver, able to deal with issues that had previously evaded human brainpower, revolutionise industries by disentangling production from human labour, lead to superior human capacities, including through potential merger of humans with AI, and generally becoming the ordering force of human societies to come (see e.g. Hoque, 2023).

Simultaneously, on the nostalgic side, the (existential) threats discourse is prominent, focusing on a broad range of possible effects of AGI, including negligent or intentional development of unsafe AI, inadequate management of technology, 'the replacement of human work, political and military interference, malicious use of AGI, wealth concentration in AGI owners, human loss of purpose and meaning, and the removal of humans as the dominant species' (Salmon et al., 2023: 376, see also McLean et al., 2023), with the latter two particularly pertaining to anthropocentric sensibilities. Nevertheless, some of the doomsday scenarios appear to be paradoxical in themselves: while the AI capable of causing an existential threat to humans is expected to be superintelligent, that is, having a cognitive capacity by orders of magnitude exceeding that of humans, it is simultaneously expected to be just as dumb and lacking in common sense as current narrow AI, operating under the same narrow efficiency calculus (see e.g. Salmon et al., 2023). This is the central premise of e.g. the paperclip thought experiment (see Bostrom, 2014), whereby AI, tasked with maximising the efficiency of producing paperclips ultimately ends up consuming the entire universe as a resource. In other words, superintelligence is expected to be as lacking in common sense as current narrow AI is and, likewise, manifesting the exploitative tendencies of humans. Likewise, scenarios of rogue AI that has no care of human life due to human cognitive inferiority assume that superintelligent AI will develop attitudes identical to human anthropocentrism (that is, become AI-centric) and treat its environment with the same disdain as humans (at least in the West) have a track record of doing. In other words, superintelligent AI is deemed to be anthropomorphic in the worst sense possible. While this is not to say that the negative externalities of superintelligence (if it is to come) are to be simply dismissed, the internal contradictions of such future visions are also to be taken account of.

Overall, however, an inflection towards progressivism can be seen. The promise of AGI is presented in terms of 'a dramatically more prosperous future' if only the risks are mitigated right; moreover, even the risks are seen to fade in comparison to the benefits: '[t]oday's systems will create tremendous value in the world and, while they do have risks, the level of those risks feel commensurate with other Internet technologies and society's likely approaches seem appropriate' (characteristically of industry insider discourse, see Altman et al., 2023). Humans are seen as having the power to shape and control the direction of AI's development, particularly through regulatory measures and other governance frameworks, including proposals for international treaties to govern AI (see e.g. Carayanis & Draper, 2023). Here, again, humans have the shaping power, holding their own fate, the fate of their inventions, and the course of history in their hands, thereby at least partly occluding the challenge to human primacy. Even more importantly, though, *specific* humans (governing elites and technology company bosses) are seen as having the shaping power, pertaining to the discussion of anthropocentrism as an elite affair below.

2.3 Posthumanist Critique of Anthropocentrism

In order to better understand the nature of taken for granted assumptions about the relationship between humans and their environment (and artefacts contained therein), it is useful to turn to posthumanist critiques of Enlightenment legacy. While internally diverse, posthumanist approaches are united in their opposition to 'foundational anthropocentric assumptions underpinning Western philosophy and science whereby the rational and intentional human subject is seen as the locus of epistemological and moral agency, responsibility, and accountability' (Mauthner, 2019: 679; see also Coeckelbergh, 2020: 40; Susen, 2022: 64) or as a 'center for orientation' (Thomsen & Wamberg, 2022: 1). Posthumanism can thus be defined as 'an ethical and political stance that promises to take seriously the problem of anthropocentrism and its deconstruction' (Herbrechter et al., 2022: 5). In this context, explicit rejection of any attempts to give preference to humans vis-à-vis other entities on rationality or any other grounds should be seen as the crucial feature of posthumanism (Nath & Manna, 2023). Such a transformation would also necessitate the capacity to think outside and beyond notions of human control and human-focused goals and values (Mellamphy, 2021: 23–24). A broader question must also be asked: 'Do you really know what it means to be human?' (Herbrechter,

2013: 38). Crucially, the definition of the human has never been stable as many of those currently bearing the label had not been considered as (fully) human while some *still* lack complete and adequate recognition; hence, even the very 'human' in human-centrism cannot be taken as a given (Braidotti, 2013: 1), let alone any hierarchies elevating this construct above the rest.

With the above in mind, ethical value is derived from understanding the world in terms of non-hierarchical and non-linear assemblages (DeLanda, 2016) or agglomerations (Kalpokas, 2021) in which diverse elements come together in an interactive fashion—a 'lively web of interdependencies' (Cielemęcka & Daigle, 2019: 81). Hence, in posthumanist and adjacent critiques of anthropocentrism, the notion of independent and self-sufficient human agency driven by supposedly detached cognition as a necessary cause for anything to truly happen is not only no longer applicable (Monforte, 2018: 380) but, instead, embodiment, embeddedness, and relationality are understood to be the natural condition that precludes appeals to universality. Hence, the adoption of such a distributed relational view undermines the possibility of a hierarchical independently accessible structure of truth. After all, if the self is decentralised, it likewise cannot be the autonomous subject that surveys the world in a detached (supposedly objective) fashion. Instead, what individuals must be understood to share is a *groundless* existence—one in which any sense of the self and the world and any political agglomerations based on the latter are suspended within a moment of tension and undecidedness.

As stressed by Susen (2022: 64), 'the binary distinction *human/non-human* has been foundational for European thought since the Enlightenment.' It is, therefore, this same binary that can ultimately be seen to undergird the taken for granted assumptions challenged by digital technologies, particularly since the elements of the aforementioned distinction have not been seen in equal terms: it was always the human who held a position superior to the non-human. Indeed, as Ferrante and Sartori (2016: 176) assert, anthropocentrism can reasonably be seen as 'the main paradigm of thought in western society' through asserting human supremacy on at least three levels that have effectively become part and parcel of the overall cultural fabric: namely, that 'humans are special and privileged entities compared to other living beings (ontology), they are the only sources of knowledge (epistemology) and the sole holders of moral value (ethics)'. Hence, the anthropocentric view presupposes that 'as beings who have unique and exceptional qualities, humans deserve a standard of care that exceeds that of other beings' (Srinivasan & Kasturirangan, 2016: 127). Such views are also reflected in legal concepts, such as human exclusivity in ownership of (almost) all things natural and artificial (Burdon, 2010: 63; see also Käll, 2023)—even in the case of property-owning legal fictions (non-natural persons, such as companies), these are entities not only established but also governed and represented by humans for human benefit. In lieu of such assumptions, posthumanism postulates the need to refuse 'ontological priority, primacy, superiority, or separateness' to any entity, humans included (Mauthner, 2019: 679). Against the symbolic and material violence of hierarchical anthropocentrism, posthumanists argue for a de-centred ethics of horizontal inclusion in which all summands carry equal value (Coeckelbergh, 2020: 42; see also Ferrante & Sartori, 2016: 177).

Grear (2015), meanwhile, makes an important observation: even the terminology (as in anthropocentrism or human-centrism) is imprecise, shaped by both historical and present patterns of discrimination and subjection: after all, the 'human' (*anthropos*) that lies at the very heart of anthropocentrism only refers to nothing more than the privileged segment of humanity—essentially, an affluent white Western heterosexual able-bodied cis-male (see also e.g. De Lucia, 2017: 189). Crucially, the ability of this minority to allow themselves the imagined luxury of Cartesian disembodied detachment has always come at the expense of non-privileged humans and the entirety of non-human environment, which today also includes digital artefacts of every kind (see also Kalpokiene & Kalpokas, 2023). It is, therefore, only natural that Grear (2015: 231–232) refers to the anthropos benchmark as 'a thoroughly gendered, raced construction' which produces 'marginalised, colonialised "others"' and renders them 'susceptible to acts and forms of domination legitimated by assumptions concerning (white) masculine superiority and mastery' is particularly powerful. It is, nevertheless, important to stress that nothing in the preceding implies that anthropocentrism would become a justifiable stance if only inter-human hierarchies were eliminated—the posthumanist goal, instead, is to eliminate all hierarchies, within and outside the 'human'.

It should not come as a surprise that posthumanism made one of its first appearances in environmental studies—after all, the unfolding climate catastrophe clearly lays bare the hubristic dangers of anthropocentrism. Likewise, the various development-focused studies, even if they are coated with a sustainability focus, manifest a strong anthropocentric perspective by envisaging 'a "good" human life as one that is freed from the vicissitudes—the risks and vulnerabilities—of living on the planet, of being part of "nature", of being animal', overcoming the risks and challenges of being placed on this planet and embedded in the physical and biological nature (Srinivasan & Kasturirangan, 2016: 126). In this way, insulated and comfortable *human* life remains the projected driving force, ensuring continued domination, even if in a 'sustainable' way—which, to decode the lofty vocabulary, simply means adjusting human domination and exploitation so that there still remains something for the future generations to dominate (see, generally, Srinivasan & Kasturirangan, 2016). Similarly, as Grear (2015: 227) asserts, the 'anthropocentrism' of modern law, by being structured around a very specific type of privileged *anthropos*, leads to *both* lack or climate crisis response and environmental degradation on the one hand *and* 'structural disempowerment [of] vast and growing numbers of human beings' on the other. Even when the environment (and it is still exclusively *natural* environment, with no mentioning of the digital one) *is* considered, that is typically done not for its own sake, but only because, and to the extent to which, it is deemed valuable to humans; in fact, even in some matters that are at a direct intersection between nature and human well-being, such as matters of health, the environment often appears as merely a side-note (Lawrence, 2019). Worse still, modern environmental law appears to entrench domination practices and 'epistemologies of mastery' that suggest taking charge of the environment as the right thing to do (De Lucia, 2017: 183). Likewise, as stressed by De Lucia (2017: 190), references to 'rights' discourse, even in terms of rights *of* environment (not to be confused to the deeply problematic *human* right *to* environment) leave too many unanswered questions as to the normalised referent of rights, in whose example are the rights constructed and with whom the power of conferring rights lies, thereby retaining the 'anthropocentric frame of reference' (similar issues also pertain to robot rights—see Gunkel, 2018). Here, again, anthropocentrism becomes manifest, on this occasion—by humans self-arrogating the right of 'conferring or extending of rights to others as a kind of benevolent gesture or even an act of compassion', thereby ignoring the human lack of self-mastery (Gunkel, 2018: 175).

Crucially, support for this point of view also goes beyond normative philosophical debates but can also be found in neuroscientific research. For example, as Oliver (2020: 136) contends, '[t]he idea that we are independent, discrete entities is a falsehood created by evolution and exaggerated by our modern culture'. The problem thus diagnosed is that 'we suffer from egocentrism—the illusionary belief that we are a coherent self at the centre of a universe' in lieu of acceptance of being 'fully enmeshed within it' (Oliver, 2020: 241). Similarly, for Charter (2019: 5), a deeper look into the functioning of the human cognitive apparatus reveals that '[t]he truth is not that the depths are empty, or even shallow, but that the surface is all there is'. Instead of a stable autonomous self, the latter is being invented on the go as '[o]ur mind is continually interpreting, justifying and making sense of our own behaviour, just as we make sense of the behaviour of the people around us', which leads to the mind being best understood as 'a consummate improviser, inventing actions, and beliefs, and desires to explain those actions, with wonderful fluidity' but without a stable underlying essence (Charter, 2019: 6–7). Again, this point of view would preclude the existence of a stable universal truth, equally available to all at all times. What *is taken* as truth, instead, pertains to an interpretation of interactions of the subject with their surroundings. Context- and subject-independence therefore cannot be assumed.

2.4 Recasting the Cartesian View

By contrast, the Cartesian point of view, foundational to modern Western thought, is predicated upon 'a conviction that the world can be elucidated in causal terms, that knowledge is about prediction and control, and that there is no limit to what men can achieve provided they have the will and the knowledge' (Dewandre, 2020: 4). To make the preceding possible, the world has to be transparent and clear-cut, since that is necessary for individuals to maintain control and autonomy (Dewandre, 2020: 4). In short, the human gaze must be able to see through any happenings of this world or, even better, the world needs to be (re)organised so that it is comfortable and convenient to the human gaze. Coextensively, an aspirational human existence is seen as one detached from not only dependence on but even being part of nature, with the very idea of development being premised upon the breaking of that bond (Srinivasan & Kasturirangan, 2016: 126), manifesting an expectation of an exceptional human standard of care and priority value ascribed to humans, with all the rest being reduced to instruments for the achievement of human wellbeing (Srinivasan & Kasturirangan, 2016: 126–127; see also Ferrante & Sartori, 2016: 176). Humans would then be seen as standard-bearers and singular value-determining actors who have self-arrogated the monopoly to, from their privileged position, ascribe status (and any rights associated therewith) to other entities (Celermajer et al., 2021). A necessary counterpart to such thinking has been the strong separation between humans and nature but also between 'reason' and 'nature' *within* the human, with an exalted version of the cultivated detached reasonable person contrasted to supposedly more 'primitive' ways of human existence (with all the discriminative, racist, and colonial effects that this separation has had and continues to have). Unsurprisingly, as Yusoff (2021: 665) contends, 'the birth of humanism is also the birth of Europe's colonial project'. Indeed, not only such accounts continue the tendency to re-render the world in accordance with a singular (and very particular) standard of the human, but also they have the effect of recasting alternatives as illegitimate non-options.

Consequently, instead of separation and dualisms (human/nature, mind/body, reason/emotion etc.) that are prevalent in Cartesian thinking but seen by posthumanists as entrenching discrimination, exploitation, and violence, the advocated stance posits interconnectedness and monism as defining feature of any existence, thereby making it impossible to establish any hierarchy of meaning and value (Margulies & Bersaglio, 2018: 103). Instead, reality is reimagined in non-hierarchical terms, with the focus shifting on the ways in which life (and other forms of existence) is seen as 'an ongoing composition' (Neyland, 2019: 11), characterised by 'the process of collectively rearranging relations, actors, and collectives' (Hoppe, 2020: 138). Likewise, for Celermajer et al. (2021: 123), 'a core feature of posthumanist approaches is the recognition of human beings' inextricable embeddedness in biological and technological worlds' or, in the words of Braidotti (2019: 45), the key consideration is that '[w]e are relational beings, defined by the capacity to affect and be affected'. Under such conditions, it becomes clear that external, interaction-independent criteria and anchoring points for discourse, including truth-claims, would be impossible. Instead, any (unavoidably fluid, as a result of interactivity) subject positions would only arise from permanently embedded simultaneous relations between multiple entities.

Appropriately, Kotzé (2020) argues for the inclusion of different 'spheres'—from biosphere and atmosphere to 'technosphere'). Consequently, a biological-physical-digital interplay—a 'socio-technico-natural compound' (Dewandre, 2020: 3)—ought to be seen as the norm to the extent that that boundaries between 'meaning and matter, the social and the technological' are never stable (Gherardi, 2012: 40). Notably, as Mahon (2018: 80–81) asserts, 'AI is explicitly posthuman because it actively erodes the distinctions between machines, animals and humans' by way of its technological (re)enactment of intelligence. It comes as no surprise that, from an anthropocentric point of view, the realisation of there being other entities with qualities and capacities previously held to be uniquely human is a traumatic one. After all, as Dewandre (2020: 4) stresses, the modern outlook is based on knowledge, prediction, and control, the loss of which can only be seen as threatening as it undermines the very premises of this deeply engrained understanding of what it means for a human to be in the world. As per the above, one of posthumanism's major contributions is, therefore, openness to the disintegration of binary thinking (as in e.g. human vs non-human), which has long dominated Western thought—and in many ways still continues to do so (Susen, 2022: 64; see also Margulies & Bersaglio, 2018: 103), clearly implying that the posthumanist project is far from complete: instead, it is perhaps best seen as still only gathering pace. From a posthumanist perspective, therefore, it is embeddedness and participation in agentic and affective flows and exchanges that characterises any entity, a human person included (see e.g. Braidotti, 2019).

Still, posthumanism as well can easily acquire progressivist overtones, for example, in aiming to conceptualise 'what it means to be no longer (quite) human (at least in a Western humanist, anthropocentric sense)' (Herbrechter et al., 2022: 12). While the difference is relatively subtle, it is evident that the implication here is that humans *had* once been the ideal subjects of anthropocentrism but have now lost this position in the face of technology. Similarly, Hayles (2022) presents a first-world focused perspective on posthumanism that equates the problematisation with humanism with technological development, thereby equating technology with moral and ethical progress. To that extent, the account in question shares its central premise with the nostalgic discourse and it is only the normative evaluation of such dislodging that differs, with progressivist posthumanists proclaiming an improvement to the understanding of human life. However, as will be further highlighted later in this book, the change should be seen not as ontological but as epistemological: on the ontological level, the idea of human independence *has always been* false, regardless of the presence or absence of technological innovation; instead, it is an epistemological shift that is taking place in terms of humans being able to acknowledge their situatedness and embeddedness which had always been the objective characteristics of the human condition. In other words, as Lupton, Clark and Southerton (2022: 363) observe, the human condition has never been purely human but, instead, 'entangled and relational with things and places'.

Some other elements of the posthumanist discourse should be taken with some scepticism as well, such as a pronounced questioning of human exceptionality by and through technology and, in particular, AI. The latter is understood as challenging and blurring the boundaries of what has traditionally been seen as specifically human (Mahon, 2018: 80–81). Nevertheless, the crucial point is that, despite such emphasis on technological novelty, one should not give in to epochal thinking: it is not the case that technology has somehow displaced or become enmeshed with humans and, therefore, rendered them posthuman but, instead, technological challenges to anthropocentrism have become *yet another symptom* that thinking in terms of human exceptionality is wrong—*and has always been*. Technology-centric thinking still plays into the human exceptionality narrative by implying the need for something equally as prominent as the human—emphatically called artificial intelligence—to joust with the human for primacy. What remains forgotten, however, is the ways in which the human is, and always has been, coextensive with and dependent upon even the simplest tools and nature's own affordances to survive. Likewise, while it is increasingly common to think of human-technology interaction through the lens of the rhetorical figure of 'the cyborg' (Miller, 2020: 269; original iterations can, however be traced back to Haraway, 1991), it must be argued that, both in abstract normative terms and, with regard to this article, specifically in terms of the role of technology in information acquisition, such thinking in itself relies on discriminatory premises.

Indeed, references to cyborgisation are problematic for several reasons. First, they render posthumanism vulnerable to misinterpretations of it being merely a diagnostic theory of the merger of humans and machines and one that uncritically accepts technological change (Osborne & Rose, 2024). The preceding even opens up an interpretation of human-technological interdependence and interactivity as 'augmentation' (Osborne & Rose, 2023), that is, as a positive contribution. Instead, augmentation, as a progressivist term, pertains more closely to one of the two forms of escapism discussed in this book. Second, along with the broader technology focus of posthumanism, there is another, more significant, avenue for criticism: one that pinpoints continuation of what could be called 'ratio-centrism' that has, among other consequences, legitimised colonial domination, that is, the assumption of the need for technological capacity to be in line with the human—or, in this case, the posthuman (Dekeyser, 2023).

To sum up, then, the posthumanist project is geared towards uncovering and manifesting the *relational* self instead of the rational and autonomous subject of Western modernity, positing the value of embeddedness and interdependence against the established drive towards transparency and, therefore, control (Dewandre, 2020: 13). Likewise, once one reframes cognition, along with agency (and, in effect, existence), as embodied, embedded, and situated, the rational autonomously self-determining subject loses all meaning (Heras-Escribano, 2019: 2; see also Damasio, 2018), leaving room for a non-essentialist encounter with the world. Hence, as argued by Charter (2019: 7), '[t]he secret of our minds lies not in supposed hidden depths'; it is, instead, 'in our remarkable ability to creatively improvise our present, on the theme of our past'. What is taken to be true (even self-evidently so) by individuals and, by extension, groups would then be yet another of such improvisations, without claims to universal validity and authority.

To conclude, then, technology performs a mirroring function in the sense that it reflects back the ultimate impossibility of living up to the Cartesian anthropocentric ideal. Instead of supposed mastery whereby the human acts as a measure of all things, that same figure emerges as a relational being intertwined with both natural and technological artefacts. Of course, technology had not *caused* this lack of mastery, but has merely exposed it (hence the mirroring). Nevertheless, this encounter with an undesired reflection gives rise to coping strategies discussed subsequently in this book.

References

Altman, S., Brockman, G., & Sutskever, I. (2023, May 22). Governance of Superintelligence. *Open AI Blog*. https://openai.com/blog/governance-of-superintelligence

Baum, J., & Villasenor, J. (2023, July 18). *How Close Are We to AI that Surpasses Human Intelligence?* The Brookings Institution. https://www.brookings.edu/articles/how-close-are-we-to-ai-that-surpasses-human-intelligence/

Bostrom, N. (2014). *Superintelligence: Paths, Dangers, Strategies.* Oxford University Press.

Braidotti, R. (2013). *The Posthuman.* Polity Press.

Braidotti, R. (2019). *Posthuman Knowledge.* Polity.

Burdon, P. (2010). Wild Law: The Philosophy of Earth Jurisprudence. *Alternative Law Journal, 35*(2), 62–65.

Carayanis, E. G., & Draper, J. (2023). Optimising Peace Through a Universal Global Peace Treaty to Constrain the Risk of War from a Militarised Artificial Superintelligence. *AI & Society, 38*, 2679–2692.

Celermajer, D., et al. (2021). Multispecies Justice: Theories, Challenges, and a Research Agenda for Environmental Politics. *Environmental Politics, 30*(1-2), 119–140.

Charter, N. (2019). *The Mind Is Flat: The Illusion of Mental Depth and the Improvised Mind*. Yale University Press.

Cielemęcka, O., & Daigle, C. (2019). Posthuman Sustainability: An Ethos for our Anthropocenic Future. *Theory, Culture & Society, 36*(7-8), 67–87.

Coeckelbergh, M. (2020). *AI Ethics*. The MIT Press.

Damasio, A. (2018). *The Strange Order of Things: Life, Feeling, and the Making of Cultures*. Pantheon Books.

Dekeyser, T. (2023). Rethinking Posthumanist Subjectivity: Technology as Ontological Murder in European Colonialism. *Theory, Culture & Society*. https://doi.org/10.1177/02632764231178482

DeLanda, M. (2016). *Assemblage Theory*. Edinburgh University Press.

De Lucia, V. (2017). Beyond Anthropocentrism and Ecocentrism: A Biopolitical Reading of Environmental Law. *Journal of Human Rights and the Environment, 8*(2), 181–202.

Denhert, M., & Gunkel, D. J. (2023). Beyond Ownership: Human-Robot Relationships Between Property and Personhood. *Social Media & Society*. https://doi.org/10.1177/14614448231189260

Dewandre, N. (2020). Big Data: From Modern Fears to Enlightened and Vigilant Embrace of New Beginnings. *Big Data & Society*. https://doi.org/10.1177/2053951720936708

Fazi, M. B. (2020). Beyond Human: Deep Learning, Explainability and Representation. *Theory, Culture & Society*. https://doi.org/10.1177/0263276420966386

Ferrante, A., & Sartori, D. (2016). From Anthropocentrism to Post-humanism in the Educational Debate. *Relations, 4*(2), 175–194.

Friedman, D. (2024). Copyright as Affirmative Action for Human Authors Until the Singularity. *GRUR International, 73*(1), 1–2.

Gabriel, I. (2020). Artificial Intelligence, Values, and Alignment. *Minds and Machines, 30*, 411–437.

Gardner, D. L., & Tanenbaum, T. J. (2023). The Access Control Double Bind: How Everyday Interfaces Regulate Access and Privacy, Enable Surveillance, and Enforce Identity. *Convergence: The International Journal of Research into New Media Technologies*. https://doi.org/10.1177/13548565231193706

Gherardi, S. (2012). *How to Conduct a Practice-Based Study: Problems and Methods*. Edward Elgar.
Graham, R. (2022). Discourse Analysis of Academic Debate of Ethics for AGI. *AI & Society, 37*, 1519–1532.
Grear, A. (2015). Deconstructing *Anthropos*: A Critical Legal Reflection on 'Anthropocentric' Law and Anthropocene 'Humanity'. *Law and Critique, 26*, 225–249.
Gunkel, D. J. (2018). *Robot Rights*. The MIT Press.
Hamzelou, J. (2023, May 25). A Brain Implant Changed Her Life. Then It Was Removed Against Her Will. *MIT Technology Review*. https://www.technologyreview.com/2023/05/25/1073634/brain-implant-removed-against-her-will/
Haraway, D. (1991). *Simians, Cyborgs and Women: The Reinvention of Nature*. Free Association Books.
Hayles, N. K. (2022). Ethics for Cognitive Assemblages: Who's in Charge Here? In S. Herbrechter et al. (Eds.), *Palgrave Handbook of Critical Posthumanism* (pp. 1195–1223). Palgrave Macmillan.
Heras-Escribano, M. (2019). *The Philosophy of Affordances*. Palgrave Macmillan.
Herbrechter, S. (2013). *Posthumanism: A Critical Analysis*. Bloomsbury.
Herbrechter, S., et al. (2022). Critical Posthumanism: An Overview. In S. Herbrechter et al. (Eds.), *Palgrave Handbook of Critical Posthumanism* (pp. 3–26). Palgrave Macmillan.
Hoppe, K. (2020). Responding and Composing: Towards a Post-Anthropocentric, Feminist Ethics for the Anthropocene. *Distinktion. Journal of Social Theory, 21*(2), 125–142.
Hoque, M. (2023, September 24). Understanding Artificial General Intelligence: The Frontier of AI. *Medium*. https://medium.com/@minh.hoque/understanding-artificial-general-intelligence-the-frontier-of-ai-de58228dbdde
Horvatić, D., & Lipić, T. (2021). Human-Centric AI: The Symbiosis of Human and Artificial Intelligence. *Entropy, 23*, 1–5.
Jacobsen, B. N. (2024). The Logic of the Synthetic Supplement in Algorithmic Societies. *Theory, Culture & Society*. https://doi.org/10.1177/02632764231225768
Jensen, T. (2023, March 9). An AI 'Sexbot' Fed My Hidden Desires – and then Refused to Play. *Wired*. https://www.wired.com/story/replika-chatbot-sexuality-ai/
Käll, J. (2023). *Posthuman Property and Law: Commodification and Control through Information, Smart Spaces and Artificial Intelligence*. Routledge.
Kalpokas, I. (2021). *Malleable, Digital, and Posthuman: A Permanently Beta Life*. Emerald.
Kalpokiene, J., & Kalpokas, I. (2023). Creative Encounters of a Posthuman Kind: Anthropocentric Law, Artificial Intelligence, and Art. *Technology in Society, 72*. https://doi.org/10.1016/j.techsoc.2023.102197

Kopelman, S., & Frosh, P. (2023). The 'Algorithmic *as if*': Computational Resurrection and the Animation of the Dead in *Deep Nostalgia. New Media & Society*. https://doi.org/10.1177/14614448231210268

Kotzé, L. J. (2020). Earth System Law for the Anthropocene: Rethinking Environmental Law Alongside the Earth System Metaphor. *Transnational Legal Theory, 11*(1-2), 75–104.

Lawrence, M. (2019). Anthropocentrism Does Not fit: Proofs from History, Science, and Logic for Ecocentric Attorneys. *Journal of Animal and Environmental Law, 11*(1), 16–27.

Lind, M., & Dickel, S. (2023). Speaking, but Having No Voice. Negotiating Agency in Advertisements for Intelligent Personal Assistants. *Convergence*. https://doi.org/10.1177/13548565231192100

Lupton, D., Clark, M., & Southerton, C. (2022). Digitized and Datafied Embodiment: A More-than-Human Approach. In S. Herbrechter et al. (Eds.), *Palgrave Handbook of Critical Posthumanism* (pp. 361–383). Palgrave Macmillan.

Magee, L., Arora, V., & Munn, L. (2023). Structured Like a Language Model: Analysing AI as an Automated Agent. *Big Data & Society*. https://doi.org/10.1177/20539517231210273

Mahon, P. (2018). *Posthumanism: A Guide for the Perplexed*. Bloomsbury.

Margulies, J. D., & Bersaglio, B. (2018). Furthering Post-Human Political Ecologies. *Geoforum, 94*, 103–106.

Mauthner, N. S. (2019). Toward a Posthumanist Ethics of Qualitative Research in a Big Data Era. *American Behavioral Scientist, 63*(6), 669–698.

McLean, S., et al. (2023). The Risks Associated with Artificial General Intelligence: A Systematic Review. *Journal of Experimental & Theoretical Artificial Intelligence, 35*(5), 649–663.

Mellamphy, N. B. (2021). Humans "In the Loop"? Human-Centrism, Posthumanism, and AI. *Nature and Culture, 16*(1), 11–27.

Miller, V. (2020). *Understanding Digital Culture* (2nd ed.). Sage.

Mitchell, M. (2024). Debates on the Nature of Artificial General Intelligence. *Science, 383*(6689), 10.1126/science.ado7069.

Monforte, J. (2018). What Is New for New Materialism for a Newcomer. *Qualitative Research in Sport, Exercise and Health, 10*(3), 378–390.

Nath, R., & Manna, R. (2023). From Posthumanism to Ethics of Artificial Intelligence. *AI & Society, 38*, 185–196.

Neyland, D. (2019). *The Everyday Life of an Algorithm*. Palgrave Macmillan.

Oliver, T. (2020). *The Self Delusion: The Surprising Science of How We Are Connected and why that Matters*. Weidenfeld and Nicolson.

Osborne, T., & Rose, N. (2024). Against Posthumanism: Notes Towards an Ethnopolitics of Personhood. *Theory, Culture & Society, 41*(1), 3–21.

Peeters, M. M. M., et al. (2021). Hybrid Collective Intelligence in a Human-AI Society. *AI & Society, 36*, 217–338.

Puntoni, S., et al. (2021). Consumers and Artificial Intelligence: An Experiential Perspective. *Journal of Marketing, 85*(1), 131–151.
Ramati, I. (2024). Algorithmic Ventriloquism: The Contested State of Voice in AI Speech Generators. *Social Media + Society*. https://doi.org/10.1177/20563051231224401
Salmon, P. M., et al. (2023). Managing the Risks of Artificial General Intelligence: A Human Factors and Ergonomics Perspective. *Human Factors and Ergonomics in Manufacturing & Service Industries, 33*(5), 366–378.
Srinivasan, K., & Kasturirangan, R. (2016). Political Ecology, Development and Human Exceptionalism. *Geoforum, 75*, 125–128.
Susen, S. (2022). Reflections on the (Post-)Human Condition: Towards New Forms of Engagement with the World? *Social Epistemology, 36*(1), 63–94.
Taylor, J. (2023, July 21). Uncharted Territory: Do AI Girlfriend Apps Promote Unhealthy Expectations for Human Relationships? *The Guardian*. https://www.theguardian.com/technology/2023/jul/22/ai-girlfriend-chatbot-apps-unhealthy-chatgpt
Thomsen, M. R., & Wamberg, J. (2022). Introduction. In M. R. Thomsen & J. Wamberg (Eds.), *The Bloomsbury Handbook of Posthumanism* (pp. 1–10). Bloomsbury Academic.
Toews, R. (2021, February 15). Artificial Intelligence and the End of Work. *Forbes*. https://www.forbes.com/sites/robtoews/2021/02/15/artificial-intelligence-and-the-end-of-work/?sh=2e90907056e3
Turner, J. (2019). *Robot Rules: Regulating Artificial Intelligence*. Palgrave Macmillan.
Xia, H., et al. (2024). The Impact of Human-Chatbot Interaction on Human-Human Interaction: A Substitution or Complementary Effect. *International Journal of Human-Computer Interaction*. https://doi.org/10.1080/10447318.2024.2305985
Yusoff, K. (2021). The Inhumanities. *Annals of the American Association of Geographers, 111*(3), 663–676.
Zawieska, K. (2020). Disengagement with Ethics in Robotics as a Tacit Form of Dehumanisation. *AI & SOCIETY, 35*, 869–883.

CHAPTER 3

Reality Nostalgia: Musings on the Post-Information, Post-Truth Society

Abstract Post-truth tends to be framed as an aberration—a supposed loss of human capacity to adhere to truth and factuality, with core decision-making criteria allegedly being replaced by emotionality, biases, and opinion-congruence. The effects of technological change, such as cognitive overload and algorithmic content governance, are conflated with a postulated increase in human vulnerability and gullibility to produce a toxic mix of what is seen as complete disregard to facts and traditional criteria for veracity. Against this nostalgic narrative of supposedly lost rationality and autonomy, this chapter redirects attention towards the primacy of narratives and other sense-making structures over strict factuality. It transpires that what post-truth, with its technological underpinnings, actually signifies is not a loss of truth but, instead, an unmasking of Cartesian fiction.

Keywords Factuality • Rationality • Disinformation • Fake news • Narrative

One of the clearest examples of nostalgic escapism with regard to digital technologies can be found in the mainstream discourse on post-truth as largely technologically induced disintegration of not only the public sphere

I. Kalpokas, *Technological Governance and Escapism in Times of Accelerated Change*, Information Technology and Global Governance, https://doi.org/10.1007/978-3-031-60890-2_3

but, even more fundamentally, human capacities crucial to anthropocentrism, such as independent reason. In this case, a noteworthy and illustrative feature in terms of how the perception of and relationship with digital communication technologies has changed has been the change from emphasising what was initially supposed to be an 'information age' towards fears of a 'post-truth era'. Indeed, the initial discussions and predictions were mainly focused on the emergence of 'a culture organized around electronic media, including [...] computer-mediated communication networks', leading to a new form of society—a networked one (Castells, 1997: 10). Information was universally seen as 'the defining feature' of the times (Webster, 2006: 263). While there is a marked continuity when comparing this information-focused outlook to the current focus on *data* as the cornerstone of society, business, and other spheres of life, there is a correspondingly clear difference between *information* as a readily available public good belonging to some form of global commons and *data* as a resource that has to be extracted and analysed at great computational cost (and is typically rendered proprietary in the process). Also, while information was seen as emancipatory by definition and synonymous with truth, data, as shown below, is largely presented as enabling deception through personalisation and, therefore, constitutive of post-truth.

In a similar fashion, the internet itself was originally itself seen as a 'liberation technology' whereby networks of connected citizens would be able to self-organise for greater democracy and freedom (see, notably, Diamond, 2010). Similarly, according to Pfister and Godana (2012: 2), the internet, being 'more decentralized and with a lower cost of entry than the mass media' was seen as being able to uphold 'the kind of lateral, many-to-many communication often theorized as the basis for democratic practice' (for an earlier argument along the same lines, see e.g. Balkin, 2004: 7). Nevertheless, as shown below, the same decentralised structure of information flow has now been incorporated in the dominant nostalgic theory of post-truth as pertaining to the loss of the traditional hierarchies of information authority and, therefore, 'truth'.

3.1 Post-Truth and Changes in the Information Infrastructures

As summarised by Harambam, Grusauskaite and de Wildt (2022: 787), 'most post-truth discourse presents clear-cut distinctions between the esteemed objective realm of facts, science, and reason and the dangerous subjective realm of emotions, ideology, and irrationality'. This supposed dichotomy also represents the crux of the nostalgic narrative about the present which, allegedly, has lost any meaningful relationship with veracity. Such nostalgia is also characterised by discussions of an 'epochal shift' from truth to alternative anchorings of discourse (see e.g. Barton, 2019). To an extent, it might be claimed, with Harjuniemi (2022: 279), that one needs to question 'the moral panics and technological determinism surrounding the somewhat ahistorical discourse on post-truth and fake news'. Nevertheless, the latter assertion is also not unproblematic: while it is certainly true that understanding the 'post' in post-truth in terms of a temporal sequence is simply ahistorical, outright dismissal of the post-truth discourse as simply a moral panic is not justified either. Reframing it as a moral panic simply recasts the narrative in progressivist terms and, therefore, fails to engage with the critical potential of the idea of post-truth. This potential is seen here as pertaining to a recasting of the concerns undergirding the mainstream discourse on post-truth in terms of a mirroring relationship between technology and humans.

Characteristically, Dahlgren and Alvares (2013: 26) stresses how 'we are witnessing historical changes in the process of production of knowledge, characterised by high velocity and dizzying excess, as well as the development of new forms of digitally derived knowledge'. Moreover, it is stressed how the advent of social media in particular has disaggregated news supply: instead of competing as collective offerings (a newspaper, a news broadcast etc.), now news and other media content compete as individual items which, first of all, results in a further increased competition but also, secondly, leads to a decontextualisation of such individual items that disrupts traditional content selection criteria (Munger, 2020: 377). To this effect, Dahlgren's argument of dizzying excess seems to be confirmed, if not pushed further. The relative absence of gatekeepers in the era of social media can be seen as a further contributing factor, enabling a seemingly unlimited flow of user-generated content (McDermott, 2019: 221). In practice, of course, this flow is never unlimited due to content moderation (both human and algorithmic), although there is still no denying that the flow of content is still unprecedented.

Similarly to the above, for Vaidhyanathan (2018: 80), the overabundance of information in today's digital communication environment means that '[m]anaging and filtering [it] has become a valuable function as a result'. Couldry and Hepp (2017: 112–113) also formulate their argument in similar terms, emphasising how '[t]he challenge of communication overload is that each message *can* be heard—as the carrier of a distinct meaning—yet it cannot be attended to' due to the lack of time that could be dedicated to understanding what each content item in the ambient digital noise actually means, leading to the need for individuals to 'drastically *select from* the environment' so that attention can only be paid to what *seems* to be noteworthy. The emergence of an unprecedented amount of information generally tends to be blamed as contributing to the impoverishment of public discourse. For example, as Ecker (2018: 80) puts it, not only 'people are faced with a deluge of information' but also they lack the necessary resources, such as 'time, motivation, cognitive capacity, and access to high quality information' that would be necessary for an adequately informed public to emerge. Meanwhile, given such a lack (or a set of different kinds of lack), it becomes an efficient strategy, at least at an individual level, to evaluate the content encountered by way of asking 'how does this fit in with what I already believe' (Ecker, 2018: 80; see also Ball, 2017: 179–180). Under such circumstances, it is not difficult to understand why framing the present time as one of unprecedented polarisation based on opinion-centric preferences appears to be a straightforward (although, as argued below, only partly correct) interpretation.

Unsurprisingly, attention has become perhaps *the* keyword when discussing technological transformations of the communication environment: in the words of Valaskivi (2022: 168), 'attention is pivotal in our media environment today'. In this situation, attention (or, to be more precise, attraction thereof) becomes a crucial variable. Since attention, due to the overabundance of potential objects of its allocation, is scarce, this leads to 'an overriding push by corporations and institutions to capture, mobilize, and profit from [it]' (Doyle & Roda, 2019: 1–2). The drive to capture attention is complemented with the algorithmic logic of information delivery, particularly via social media platforms, that enables targeting and personalisation of content, that is, offering an individual what they are bound to like (and, therefore, pay attention to). The net result, however, particularly as such processes are extrapolated across a given population, is 'the growing personalization of constructed realities and the subsequent individualization effects' (Just & Latzer, 2017: 254; see also Thorson et al., 2021). In this way, reaching the 'right' people has become both an obsession in communication campaigns and a major promise of online platforms, with all the manipulation worries that such targeting potentially entails (Cotter et al., 2021). Still, it is important not to give in to the discourse of unprecedentedness: after all, certain human proclivities can be targeted because they *already* exist and not because they are 'invented' through technology.

It is also the case that the overabundance of digital information has transformed audience relationship with news and news-related content. On the one hand, 'news today is ubiquitous, pervasive, and constantly around us', often to the extent that individuals begin expecting the news to find them, and not vice versa (Gil de Zúñiga et al., 2017: 106–107; see also Park & Kaye, 2020: 19). Crucially, though, it transpires that the *quality* aspect of information has no effect on citizens' political participation: what really matters is whether they *feel* informed; hence, those driven by deficient information are just as likely to make their voices and opinions heard and actively push for opinion-commensurable political decisions as those who possess verifiably factual knowledge, thereby leading to further proliferation of a-factual points of view and their inclusion in the political agenda (Song et al., 2020: 64). On the other hand, though, people may become so overwhelmed and anxious about the ever-increasing stream of news content that they begin avoiding it altogether, thereby even further deepening their dependence on piecemeal haphazard encounters (Toff & Nielsen, 2022).

Moreover, the diversity of views, opinions, and fact-claims means that trust is severely hampered as everything and nothing seems to be potentially provable (O'Shaughnessy, 2020). In this context, it is important to note that lack of trust is typically associated with a higher likelihood of belief in fake news and conspiracy theories (Anthony & Moulding, 2019; Pierre, 2020). After all, if everything and anything (and, likewise, nothing) can equally be true, it is only tempting to turn to claims that 'stick' to already held biases, opinions, and preconceptions. Crucially, because there is so much to choose from, one can always find opinion-congruent content, regardless of the level to which it corresponds with verifiable facts (McDermott, 2019: 221). In addition to emotion- and bias-congruence, popularity is also seen as a substitute for truth and 'impersonal logic', replacing them with mere 'tribal affinity' (Hannan, 2018; for a similar argument, see also Sunstein, 2018: 74). Moreover, such selective sorting is further strengthened by the online platforms themselves, whereby content-to-be-liked is algorithmically selected and displayed to any given user. Consequently, the current transformations of the public sphere have the tendency to result in fragmentation into opinion-congruent bubbles (Seeliger & Sevignani, 2022: 11). Likewise, in the context of attitude polarisation, it is claimed, individuals tend to opt for opinion-congruent and easily processed content, with the effect of often literally seeking out disinformation (Woolley & Joseff, 2020: 6).

The proclivity towards opinion-congruence can also be abused in the form of manufacturing false unanimity, either by locking individuals into groups of the like-minded or by throwing fake accounts into the mix (Chadwick & Stanyer, 2022). Emphasis on the importance of opinion-congruence on the aptitude to believe content (particularly in situations that lack clarity or when high stakes are involved) is indeed commonplace in research on post-truth disinformation and conspiracy theories (see e.g. Davis, 2017: xix; Douglas et al., 2017: 539; Gilead et al., 2018: 7; McIntyre, 2018: 45; Bruns et al., 2020: 13; O'Shaughnessy, 2020: 64; Uscinski & Enders, 2022). Opinion-congruence is also seen as a crucial driver behind contemporary propaganda campaigns, particularly as disinformation agents are seen to be targeting diverse groups simultaneously not with the aim of hammering one centralised message (which would have been characteristic of earlier iterations of propaganda) but, instead, pushing separate narratives to such groups in order to induce the disintegration of public discourse through ensuring that the different factions no longer have anything in common to talk about (Bastos et al., 2021). This attitude also has clear undertones of a specifically negative outlook towards the public itself, implying its deteriorated interest in (or capacity for) cooperation and cohesion. Moreover, it transpires that opinion-congruence also works when factchecks and corrections of disinformation are encountered, albeit unevenly: if an issue is not particularly important or engaging, correction is likely to be internalised but if the issue is seen as one of importance and the individual in question is already highly involved in it, then corrections typically have limited or no effect (Vidigal & Jerit, 2022).

Overall, then, when discussing post-truth, it has become commonplace to assert that content abundance and the speed with which new content emerges have led to an 'epistemic crisis' (Dahlgren, 2018) that erodes civic participation and thus democracy. Post-truth is thus seen to nullify any claim to objectivity as any utterance can be deemed biased. Thus, the modern model of truth-telling appears to disintegrate as 'common ground and norms for democratic debate evaporate' (Harjuniemi, 2022: 272). Similarly, the fragmentation of the media environment is framed as implying that everybody now can immerse themselves in a reality of their choice and enabling politicians to benefit from the 'strategic extremism' of their own positions (Lewandowsky et al., 2017: 359–360). From a content dissemination and consumption perspective, then, it seems that today's digital technologies have left truth simply 'beside the point' (Vaidhyanathan, 2018: 13). In this way, a key element of technological determinism in discourse on post-truth becomes clear.

3.2 Post-Truth and the Imagery of Societal Decay

What is either implicitly or explicitly stressed in most mainstream accounts of post-truth is that a certain weakness of (moral) character seems to have taken over the general population: Schindler (2020: 392), for example, stresses how 'we live in a time when many human beings cannot realise their capacity to relate in an independent, reflexive manner to the world' as '[s]omething is driving people away from confronting the difficulty of relating to a world they do not fully control'. The preceding is then seen to lead towards 'the rejection of basic principles of reason and veracity' (Hopkin & Rosamond, 2017: 1). Moreover, the information environment is also pronounced as such that individuals are empowered 'to choose their own reality' in which belief trumps evidence (Lewandowsky et al., 2017: 361). This focus on convenient and emotionally appealing reality is also seen as causing complete disregard for any evidence to the contrary, however factual such evidence might be (Horsthemke, 2017: 275; Barton, 2019; Baier, 2024). Similarly, for Baier (2024), individuals appear to be caught within an endless loop of emotional frenzy and tribal affinity, thereby disregarding factual truth and established arbiters thereof. In effect, then, individuals are seen as losing themselves in an emotional frenzy manufactured for them by today's digital media and, therefore, losing baseline touch with reality (see e.g. Brunkhorst, 2024).

It transpires, however, that structural and individual factors are taken to mutually reinforce each other: in the words of Hopkin and Rosamond (2017: 12), simultaneously individuals are overwhelmed with low-quality information that abounds in the media environment *and* there is a corresponding diminution in their capacity to distinguish between truthful and patently false truth claims. To that effect, one can witness a determinist attitude—some individuals are simply written off as doomed to follow misleading claims while others (including, presumably, those passing judgement) are somehow immune and, therefore, deserving of a higher position in the epistemic hierarchy. Crucially, the present situation is seen as overall disintegration of traditional forms of authority, such as scientific knowledge: in the words of McIntyre (2018: 17), '[o]nce respected for the authority of its method, scientific results are now openly questioned by legions of nonexperts who happen to disagree with them'; moreover, the voices of such nonexperts get amplified on social media in a way that was never possible before (for a similar argument, see also Gordon, 2018). Statements like this clearly demonstrate the nostalgic structure of the argument being made: that of bygone times that had been characterised by a 'proper' information order that has been—temporarily or irrevocably—lost. Likewise, Enroth (2021) presents post-truth as not just an epistemic crisis but, first of all, a crisis of authority (see also de Albuquerque et al., 2022: 1200). In particular, not only politicians but also, for example, social media influencers are framed as challenging traditional information hierarchies and encouraging distrust in, for example, science and scientists (see e.g. Hassell & Chin, 2023). No less importantly, it is now increasingly stressed that some of the new authority figures may not even be humans, ranging from computer-generated influencers (Block & Lovegrove, 2021; Sands et al., 2022) to bots (Hagen et al., 2022), thereby further contributing to the framing of digital technologies as threats.

Similarly, Kavanagh and Rich (2018: x–xi) define what they call 'truth decay' as a combination of growing societal inability to agree on basic facts, increasing centrality of opinion and personal experience at the expense of detached factual knowledge, and diminishing trust in traditional knowledge authorities. Likewise, McDermott (2019: 218) emphasises what is presented as a harmful parity between opinion and facts or emotions and statistics, thereby adversely influencing 'public policy debates as well as [...] behavioral outcomes (e.g., voting)'. Authors working within the field of conspiracy theories also tend to emphasise something akin to 'a reciprocal relationship between mistrust and misinformation' (Pierre, 2020: 630), particularly inasmuch as it pertains authority structures (Oleksy et al., 2021: 2) as well as mainstream media and other traditional news sources (see e.g. Hendricks & Vestergaard, 2019: 98; Hameleers, 2020: 3). Also in the same vein, Foroughi, Gabriel and Fotaki (2019: 140) describe the present condition as '[t]he breakdown of consensus about a single all-encompassing truth', which in turn enables biased understandings of the world, including the ability to ignore unpalatable information. The perception is that, as summarised by Farrell (2017: 8), '[t]he old certainties are gone'. Notably, while the above perspectives may somewhat differ in the angles taken, reality nostalgia is clearly evident in all of them nevertheless, manifesting a somewhat conservative emphasis on 'proper' order and values having been eroded or at least appearing to be under threat.

Notably, post-truth tends to be seen as a structural transformation that fundamentally alters the very premises of public discourse: for example, as Bufacchi (2021: 349) claims, post-truth 'doesn't simply deny or question certain facts, but it aims to undermine the theoretical infrastructure that makes it possible to have a conversation about the truth'. Again, the structure of the argument is clear: while the past had been characterised by a normatively more preferable relationship to truth, the present can be seen to denote exactly the opposite, namely, the undermining of the very conditions that enable truth as a social factor. There is also, at least to some extent, a sense of elitism in contemporary takes on post-truth and conspiracy theories, presenting them, for example, as 'the "have nots" version of correlational "knowledge", whose function is not descriptive but instrumental' (Andrejevic, 2020: 33). Such a deterioration of public discourse is thus often seen as a foregone conclusion, characterised by 'the diminishing importance of anchoring political utterances in relation to verifiable facts' (Hopkin & Rosamond, 2017: 2). Such accounts can also be seen to manifest, not infrequently, more sinister, even quasi-conspiratorial, overtones, particularly in referring to 'the constant sophisticated attempts to confuse and dupe, with considerable public epistemic and legislative effects' (Harsin, 2019: 102).Similarly, for Bufacchi (2021: 354), post-truth is best understood as 'an invention of the powerful': in fact, '[i]t comes from the top, specifically with the intent to undermine the Truth' because truth, it is claimed, can only serve itself but not those in power whereas post-truth is sufficiently malleable to enable a power-knowledge nexus (for a similar argument, see also Foroughi et al., 2019: 137). For others, meanwhile, this need to abuse contemporary information flows is less of a masterplan but, instead, a necessity faced by political actors as they are now supposed to strategically manage and leverage online opinion formation processes (see e.g. Mazzoleni, 2017: 142). However, a contrast between an idealised more reasonable past and a deteriorated present is drawn in either case.

Some would even go as far as equating adherence to fact-incongruent content with an irrational attitude to the world as such, thereby further strengthening the idea of some sort of moral and cognitive downfall being underway and parts of the population are, again, doomed to adhere to post-truth. Characteristically of such attempts, for McIntyre (2018: 9), 'post-truth is when one thinks that the crowd's reaction actually *does* change the facts about a lie'. The belief that public sentiment truly has fact-altering qualities would, of course, seem nonsensical—and perhaps rightly so. A more accurate interpretation, meanwhile, would focus on the social effects of such content, thereby going *beyond* focus on facts. Post-truth is also often framed as denoting a rather cynical attitude—as 'a challenge not to the idea of *knowing* reality but to the existence of reality itself' (McIntyre, 2018: 10). However, this is very much an oversimplification: what happens instead is a plurification of the socially perceived reality each version of which is vehemently adhered to. In other words, the reason why people side with post-truth is not because they do not believe in the very existence of reality but, instead, because a particular version of reality strongly appeals (or is made to appeal) to them. Given the above framing of post-truth, it is unsurprising that critics tend to deride it as 'deplorable' (McIntyre, 2018: 9) and something to be explicitly and intentionally fought against (Bufacchi, 2021: 357). In short, the 'better' times of the past have to be brought back—in other words, the quest is to 'make Truth great again'.

A notable group dimension is involved here. On the one hand, there is a clear 'us vs them' distinction at the centre of conceptualising post-truth and vulnerability to disinformation more broadly: the rational 'us' versus the irrational 'them' as it is always the others who are seen as inadequate or simply weak. On the other hand, though, there is a fair amount of depersonalisation going on as well, with the 'them' typically being reduced to some iteration of 'the masses'. In a subtler way, this massification is also reflected in an alleged incapacity for independent rational thought and propensity to group thinking and herd instinct. Under such interpretations, individuals are seen to almost mindlessly follow their peer groups and, as if automatically, self-segregate onto communities of the like-minded (see e.g. McIntyre, 2018: 38–39; van Prooijen & Douglas, 2018; Mazarr et al., 2019: 35; Wanless & Berk, 2020: 89). While it is clear that self-selection (and, no less importantly, selection through automated content moderation, e.g. on social media), this should not be taken to infer a sudden loss of individual agency.

In addition, technological innovations in content creation tend to be seen as factors further contributing to the deterioration of the public sphere. Deepfakes could be perhaps the most prominent example in this respect, particularly in terms of creating false proof of events that did not take place or obfuscating reality more broadly by erasing the lines between reality and fantasy, proof and falsification (Woolley, 2020: 14). Particularly when combined with the opportunities to microtarget groups that are particularly vulnerable to certain messaging, for example, through social media, deepfakes are seen to be as a particularly powerful manipulation tool (Dobber et al., 2021). Deepfakes are also generally seen as actively contributing to the erosion of trust (Woolley, 2020: 107) and causing 'a significant threat to human society and democracy' (de Rancourt-Raymond & Smaili, 2022: 3) due to their alleged potential to turbocharge disinformation. In particular, the ensuing democratisation of content creation has a tendency to be seen a releasing the darker drives of the masses who no longer need significant skills to engage in misdeeds (Hight, 2022). As a consequence, then, disorientation and inability to commit to any version of events are seen to ensue (Vaccari & Chadwick, 2020: 9). The preceding, then, can be easily connected to the broader discourse on post-truth as dominance of opinions over facts, and the general deterioration of public discourse (see e.g. Whittaker et al., 2020: 95). The sentiment here can also be traced to the broader conservative fear of 'the masses' and a nostalgia for times when elite judgement (here understood as a kind of 'knowledge elite') was in charge—a sentiment that can be traced at least to Plato.

Likewise, political bots are seen to disrupt knowledge patterns by creating artificial trends and a fake sense of consensus and grassroots support or simply disorienting individuals (Keller et al., 2020; Miller, 2020: 169; Hagen et al., 2022). Such, largely automated, attempts at knowledge disruption through the creation of fake grassroots voices—typically referred to as 'astroturfing'—are then seen to allow for an even more effective enclosure of individuals within their own bubbles of opinion-congruent content (Keller et al., 2020: 257; Schoch et al., 2022: 1). In quasi-militaristic terms, '[a]stroturfing agents invade […] habitats of public discourse and exploit their openness, anonymity, and reach', thereby promoting the interests of shadowy puppet masters (Zerback & Töpfl, 2022: 401). While astroturfing is not always used to promote fake news content, there is still deception at least in terms of the identity of users and the actual levels of support for particular news agenda items, meaning that audiences are misled nevertheless (Keller et al., 2020). In this case, the simultaneously feared and loathed masses seem to be amplified with what could be called 'machinic masses'.

3.3 Is There a Way Out?

What the nostalgic approach fails to appreciate, though, is the importance of the a-rational dimension in human behaviour overall. While not necessarily (or not always) irrational, this dimension accounts for criteria used for interpreting the world that do not conform to strict coherence between utterances and verifiable facts or judgements made from a detached and unbiased subject position. Of crucial importance here is the necessity of a narrative. As Holmstrom (2015: 120) puts it, people need a narrative because it 'provides explanations', that is, 'describes the past, justifies the present, and presents a vision of the future'. Meanwhile, if an equally effective factual narrative is lacking or it is incomplete, or difficult to understand, it consequently becomes impossible to fill the gap in the general structure of meaning with verifiable information and an opportunity is created for alternative accounts to emerge, particularly if they produce a more satisfying effect (Ecker, 2018: 82). Indeed, the key thing that matters is the provision of meaning to an otherwise seemingly disorienting and disconcerting reality, even if that is achieved through immersion in conspiracy theories (de Albuquerque et al., 2022: 1199). After all, the most important thing that individuals expect from a narrative is that it provides actionable insights, regardless of its veracity (Andrejevic, 2020: 32). Hence, it is crucial to keep in mind that even fact-incongruent narratives, be it conspiracy theories or fake news, have the capacity to nevertheless 'connect people, give meaning to experienced disparities and corruption in society' (Harambam et al., 2022: 785)—grievances that often do have a factual basis and that had not yet been adequately explained, perhaps because the group in question had previously attracted insufficient attention from decision-makers, the news media, and the society as a whole. This does not mean that the reaction itself is adequate from a fact-based point of view, but it often does answer a real underlying need when other responses are unavailable.

Notably, the need and capacity to establish patterns (even in cases when there objectively are none or only spurious correlations are in place) is hardwired through evolution (Butter, 2020: 66), leading people to seek existing narratives or create them where none exist or the currently held ones are seen as inadequate—again, often for subjective reason (see e.g. van Prooijen, 2018: 41; Wood & Douglas, 2018: 250; de Albuquerque et al., 2022: 199). In particular, such behaviour provides order and predictability, however imaginary, that is nevertheless easier to act on than randomness or overbearing complexity (Douglas et al., 2017: 538; Silva et al., 2017: 425–426; Douglas & Sutton, 2018: 261; Mazarr et al., 2019: 52; Butter, 2020: 67). While, on the one hand, this could be seen as a normatively flawed coping strategy (Butter, 2020: 67) that is only concerned with 'simple recipes for explaining complex realities' (Cosentino, 2020: 85), it is nevertheless an efficient solution in situations when there is ether a deficit (Bruns et al., 2020: 13) or, on the contrary, an overflow (Kalpokas, 2018, 2020; O'Shaughnessy, 2020) of actionable information.

In fact, it transpires that fact-based interventions in order to dislodge factual narratives can even be counterproductive by making the representation of the world more complex and disorienting, thereby increasing the drive for clear-cut stories that seemingly put all things in order (Holmstrom, 2015: 121). As per the above, therefore, it can be claimed that '[t]ruth, as in a fact or piece of information, has no intrinsic value' but, instead, '[i]t is up to the narrative to create that value' (Holmstrom, 2015: 124). Very similarly, for Baron (2018: 73), the crucial variable in achieving success, particularly in the political domain, 'is not evidence (i.e. facts) but *meaning*'. Moreover, there is also a further way in which narratives, even factually false ones, can be meaningful—namely, as affirmations of identities and 'social facts', namely, 'rituals of lived experience that weave together strands of social and psychological life' (Bastos & Tuters, 2023: 3). In this sense, narratives can be seen as 'reading and partaking in stories that are not units of information travelling from sender to receiver, but a collective narrative effort', collective construction of 'a situation in which nothing new is learned, and where no information is essentially accurate, but in which a particular view of the contending forces in the world is portrayed and confirmed' (Bastos & Tuters, 2023: 4). Hence, particularly in challenging or crisis situations, when previously taken for granted assumptions are no longer effective, alternative narratives and knowledge strategies, such as conspiracy theories, abound (van Prooijen, 2018: 23; Bonetto & Arciszewski, 2021: 918).

Indeed, it must be kept in mind that comprehending any given situation and knowing how to act and respond to the changing environment are key human drives that, taken together, imply the need to resolve any uncertainty as quickly as possible—including by resorting to unverified information should other options be unavailable (Kim, 2018: 2; Schwaiger et al., 2022). For some, a crucial driver behind resorting to factually false narratives is actual or perceived powerlessness and a sense of lack of control (Douglas et al., 2017: 539; Anthony & Moulding, 2019; Butter, 2020: 74; Bonetto & Arciszewski, 2021). Since groups suffering from actual or perceived powerlessness tend to be either those who are underprivileged without a foreseeable possibility for emancipation or previously dominant groups that have been displaced (or undergoing a process of being displaced by new, more progressive, groups), their concerns are also likely to be flatly dismissed as either irrelevant or backwardly and also be assigned the stigmatising 'fake news' or 'conspiracy theory' label and, in the process, further strengthen the perception of isolation. Belonging to a community of those allegedly in the know therefore brings a sense of subjective empowerment (Cosentino, 2020: 85; see also Oleksy et al., 2021: 2). Here, a somewhat more nuanced interpretation becomes available: instead of a rather flatly stigmatising view that equates an underprivileged condition with susceptibility to disinformation, story-telling is understood in a ritualistic way, thereby regaining the status it had in traditional communities: that of collective affirmation and identity acclamation in a notably emancipatory way—a function for which factuality is, plainly speaking, irrelevant (see, generally, Bastos & Tuters, 2023).

Crucially, it must be acknowledged that ignorance is an unavoidable aspect of the human condition: after all, the world ultimately transpires to simply be too complex for individuals to make sense of—at least as rational individual cognising subjects (Sloman & Fernbach, 2017: 257). Therefore, to follow Baron (2018: 196), while it is clear that evidence still occupies a privileged place in discourse, at least as a value laden-signifier (after all, even those peddling in fake news and conspiracy theories cite what they claim to be evidence), the scope and meaning of this term necessitates expansion, not least because 'narratives are also a form of evidence', irrespective of the actual substance of such narratives. It is not inconceivable, however, at least from the nostalgic perspective, for the above argument to be turned on its head, framing the weakness of the narrative-evidence nexus as simply yet another sign of the general deterioration of the times. Nevertheless, it is argued here, the preceding is best understood as a general propensity and not as something that has emerged with, for example, the proliferation of digital technologies or with the 'era' of post-truth. Instead, such propensities have merely become more visible.

The nostalgic take on the centrality of narratives is clearly evident in the way Sloman and Fernbach (2017: 261) frame them as mere pleasurable illusions. Where they err is in failing to distinguish between different types of non-factual and a-factual narratives. For them, fictional stories that belong to the domain of the creative arts, created for entertainment purposes and offering 'no pretense of being real' are in principle identical to a-factual narratives intended to explain the social and political world (Sloman & Fernbach, 2017: 261). Meanwhile, McDermott (2019: 221) frames narratives in a distinctively emotion-laden sense while also emphasising the fictional—or, rather, mythical—dimension: according to this account, narrative and facts stand in stark opposition to each other, as if facts were able to speak for themselves as standalone items and did not need narration. Ultimately, the preceding boils down to what Harjuniemi (2022: 279) calls 'a baseless nostalgia for a by-gone era characterised by truth and reason'. There is, seemingly, a stark choice: you are either with the Enlightenment (i.e. rationality and Truth-with-a-capital-T) or with the 'charlatans' (d'Ancona, 2017: 5). This statement, however, is also revealing of the actual object that such nostalgia looks back to: the exalted Enlightenment version of the human. Likewise, when it comes to, for example, conspiracy theories, it is a common strategy to sideline them by imagining them as the domain of fringe delusional groups that are unlikely to disturb the immune majority; nevertheless, it transpires that not only conspiracy theories are recurrent throughput human history but also they result from normal cognitive processes, meaning that most, if not all, individuals are prone to at least some conspiratorial inklings (Douglas & Sutton, 2018: 257–259; see also see also van Prooijen, 2018: 17; Butter, 2020). Hence, while conspiracy theories, just like fake news, typically tend to be used as a delegitimating label directed towards others—as Butter (2020: 24) stresses, 'it is always other people who articulate "conspiracy theories"'—such a dismissal only acts as an easy and convenient distortion of reality in its own right, allowing for self-righteousness and a nostalgic reframing of the present.

Moreover, since individuals, quite naturally, are inclined to think that they and their group is firmly rooted in reality, it makes biases and false assumptions (let alone nuances of meaning) difficult to spot, instead externalising the blame to non-like-minded others (Uscinski & Enders, 2022). It transpires that concern about the threat of fake news and disinformation correlates with the perception that *others* are vulnerable: hence, the likely victim of disinformation is seen as the (allegedly, less intellectually gifted) 'other' rather than the (seemingly rational and critically minded) self (Altay & Acerbi, 2023). In other words, the very notion of 'conspiracy theory' (or fake news or post-truth, for that matter) has a tendency to function not just as a descriptor but also, and perhaps even more prominently, as a disqualifying label that not only dismisses another group's knowledge (whether or not there is a factual basis for such a dismissal is a completely different matter) but also disqualifies them as (intellectually, morally etc.) inferior (Uscinski & Enders, 2022).

In addition, the industrial aspect of fake news and disinformation is often ignored. After all, much of the fake content we encounter on social media tends to be produced in quasi-factory settings where 'falsehoods and hate are executed by competitive and collaborative teams of task-minded laborers rather than "extremist" or "fascist" believers in a coherent ideology' (Grohmann & Ong, 2024: 2). In fact, this industry transpires to be characterised by a multi-tiered structure of work, from elite fake news entrepreneurs to digital 'sweatshop' labourers; however, the same structure is also paralleled by the opposite side with the influx of funding for countering disinformation also having led to the development of factchecking-as-an-industry (Grohmann & Ong, 2024: 3–4). Hence, instead of having become uniquely fallible, audiences are often simply torn between professionalised industrial-scale processes.

It is unsurprising that Uscinski and Enders (2022) spot an 'epistemological naïveté' behind much of the current discourse on post-truth—at least the section of such literature that is overly focused on the nostalgic vision of past rationality and objectivity and the current degradation of such capacities. In particular, as shown in this chapter, the blame for the loss of the times now allegedly in the past is put on digital technologies. Nevertheless, the role of digital technologies is this case is, again, that of a mirror: they do reflect the lack of propensity for a facts-or-bust view of the world which often gets imagined onto the past. Paradoxically, it transpires that the self-righteous focus on the technologically mediated loss of access to truth might well be a subjective interpretation itself, heavily dependent on partisan cues within a particular encounter with content (see e.g. Calice & et al., 2023). As the idealised vision of the human, discussed in the previous chapter, collapses and the taken-for-granted assumptions about what it means to be a 'proper' human person become no longer sustainable, attitudes towards truth and the capacity to grasp 'reality' as such are still being held to through nostalgic escapism as a coping strategy.

Overall, instead of deserving a negative framing, post-truth does serve several important functions. First of all, it helps undermine the dominance of the Cartesian model in terms of a rational subject having a privileged access to the world. Instead of being a sign of deterioration and loss of mental and/or moral capacities, post-truth simply unmasks the underlying propensities of humans that had always been there but had been left unacknowledged. Hence, what the 'post' of post-truth does is signalling the undermining of pretence that truth and verifiable facts are, or have been, fundamental anchoring points of discourse. In this case, technology once again performs a clear mirroring function by making the a-rational ways in which humans relate to the world clearer and more visible than before. Of course, from the inherited anthropocentric position this view cannot be a palatable one, thus giving rise to a nostalgic rejection of the present.

References

Altay, S., & Acerbi, A. (2023). People Believe Misinformation is a Threat Because They Assume Others Are Gullible. *New Media & Society*. https://doi.org/10.1177/14614448231153379

Andrejevic, M. (2020). *Automated Media*. Routledge.

Anthony, A., & Moulding, R. (2019). Breaking the News: Belief in Fake News and Conspiracist Beliefs. *Australian Journal of Psychology, 71*, 154–162.

Baier, C. (2024). Narratives of Post-Truth: Lyotard and the Epistemic Fragmentation of Society. *Theory, Culture & Society, 41*(1), 95–110.

Balkin, J. M. (2004). Digital Speech and Democratic Culture: A Theory of Freedom of Expression for the Information Society. *New York University Law Review, 79*, 1–55.

Ball, J. (2017). *Post-Truth: How Bullshit Conquered the World*. Biteback Publishing.

Baron, I. Z. (2018). *How to Save Politics in a Post-truth Era*. Manchester University Press.

Barton, C. C. (2019). Critical Literacy in the Post-Truth Media Landscape. *Policy Futures in Education, 17*(8), 1024–1036.

Bastos, M., & Tuters, M. (2023). Meaningful Disinformation: Narrative Rituals and Affective Folktales. *Big Data & Society*. https://doi.org/10.1177/20539517231215361

Bastos, M., Mercea, D., & Goveia, F. (2021). Guy Next Door and Implausibly Attractive Young Women: The Visual Frames of Social Media Propaganda. *New Media & Society*. https://doi.org/10.1177/14614448211026580

Block, E., & Lovegrove, R. (2021). Discordant Storytelling, 'Honest Fakery', Identity Peddling: How Uncanny CGI Characters are Jamming Public Relations and Influencer Practices. *Public Relations Inquiry, 10*(3), 265–293.

Bonetto, E., & Arciszewski, T. (2021). The Creativity of Conspiracy Theories. *Journal of Creative Behavior, 55*(4), 916–924.
Brunkhorst, H. (2024). Cultural Industry in the Age of Post-Truth Democracy. *Philosophy and Social Criticism, 50*(1), 28–42.
Bruns, A., Harrington, S., & Hurcombe, E. (2020). 'Corona? 5G? or Both?': The Dynamics of COVID-19/5G Conspiracy Theories on Facebook. *Media International Australia, 177*(1), 12–29.
Bufacchi, V. (2021). Truth, Lies and Tweets: A Consensus Theory of Post-Truth. *Philosophy and Social Criticism, 47*(3), 347–361.
Butter, M. (2020). *The Nature of Conspiracy Theories*. Polity.
Calice, M. N., & et. Al. (2023). Polarized Platforms? How Partisanship Shapes Perceptions of 'Algorithmic News Bias'. *New Media & Society, 25*(11), 2833–2854.
Castells, M. (1997). An Introduction to the Information Age. *City, 2*(7), 6–16.
Chadwick, A., & Stanyer, J. (2022). Deception as a Bridging Concept in the Study of Disinformation, Misinformation, and Misperceptions: Toward a Holistic Framework. *Communication Theory, 32*, 1–24.
Cosentino, G. (2020). *Social Media and the Post-Truth World Order*. Palgrave Macmillan.
Cotter, K., et al. (2021). 'Reach the Right People': The Politics of 'Interests' in Facebook's Classification System for Ad Targeting. *Big Data & Society*. https://doi.org/10.1177/2053951721996046
Couldry, N., & Hepp, A. (2017). *The Mediated Construction of Reality*. Polity.
d'Ancona, M. (2017). *Post Truth: The New War on Truth and How to Fight Back*. Ebury Press.
Dahlgren, P. (2018). Media, Knowledge and Trust: The Deepening Epistemic Crisis of Democracy. *Javnost – The Public, 25*(1), 20–27.
Dahlgren, P., & Alvares, C. (2013). Political Participation in an Age of Mediatization. *Javnost – The Public, 20*(2), 47–65.
Davis, E. (2017). *Post-Truth: Why We Have Reached Peak Bullshit and What We Can Do About It*. Little, Brown.
de Albuquerque, A., et al. (2022). Coronavirus Meets the Clash of Civilizations. *Convergence: The International Journal of Research into New Media Technologies, 28*(4), 1198–1213.
de Rancourt-Raymond, A., & Smaili, N. (2022). The Unethical Use of Deepfakes. *Journal of Financial Crime*. https://doi.org/10.1108/JFC-04-2022-0090
Diamond, L. (2010). Liberation Technology. *Journal of Democracy, 21*(3), 69–83.
Dobber, T., et al. (2021). Do (Microtartgeted) Deepfakes Have Real Effects on Political Attitudes? *The International Journal of Press/Politics, 26*(1), 69–91.
Douglas, K. M., & Sutton, R. M. (2018). Why Conspiracy Theories Matter: A Social Psychological Analysis. *European Review of Social Psychology, 29*(1), 256–298.

Douglas, K. M., Sutton, R. M., & Cichocka, A. (2017). The Psychology of Conspiracy Theories. *Current Directions in Psychological Science, 26*(6), 538–542.
Doyle, W., & Roda, C. (2019). Introduction. In W. Doyle & C. Roda (Eds.), *Communication in the Era of Attention Scarcity* (pp. 1–6). Palgrave Macmillan.
Ecker, U. K. H. (2018). Why Rebuttals May Not Work: The Psychology of Misinformation. *Media Asia, 44*(2), 79–87.
Enroth, H. (2021). Crisis of Authority: The Truth of Post-Truth. *International Journal of Politics, Culture, and Society.* https://doi.org/10.1007/s10767-021-09415-6
Farrell, D. (2017). Political Science in Turbulent Times. *Political Insight, 8*(1), 7–9.
Foroughi, H., Gabriel, Y., & Fotaki, M. (2019). Leadership in a Post-Truth Era: A New Narrative Disorder? *Leadership, 15*(2), 135–151.
Gil de Zúñiga, H., Weeks, B., & Ardèvol-Abreu, A. (2017). Effects of the News-Finds-Me Perception in Communication: Social Media Use Implications for News Seeking and Learning About Politics. *Journal of Computer-Mediated Communication, 22*, 102–123.
Gilead, M., Sela, M., & Maril, A. (2018). That's My Truth: Evidence for Involuntary Opinion Confirmation. *Social Psychological and Personality Science.* https://doi.org/10.1177/1948550618762300
Gordon, M. (2018). Lying in Politics: Fake News, Alternative Facts, and the Challenges for Deliberative Civics Education. *Educational Theory, 68*(1), 49–64.
Grohmann, R., & Ong, J. C. (2024). Disinformation-for-Hire as Everyday Digital Labor: Introduction to the Special Issue. *Social Media + Society.* https://doi.org/10.1177/20563051231224723
Hagen, L., et al. (2022). Rise of the Machines? Examining the Influence of Social Bots on a Political Discussion Network. *Social Science Computer Review, 40*(2), 264–287.
Hameleers, M. (2020). They Are Selling Themselves out to the Enemy! The Content and Effects of Populist Conspiracy Theories. *International Journal of Public Opinion Research.* https://doi.org/10.1093/ijpor/edaa004
Hannan, J. (2018). Trolling Ourselves to Death? Social Media and Post-truth Politics. *European Journal of Communication, 33*(2), 214–226.
Harambam, J., Grusauskaite, K., & de Wildt, L. (2022). Poly-Truth, or the Limits of Pluralism: Popular Debates on Conspiracy Theories in a Post-Truth Era. *Public Understanding of Science, 31*(6), 784–798.
Harjuniemi, T. (2022). Post-Truth, Fake News and the Liberal 'Regime of Truth': The Double Movement Between Lippmann and Hayek. *European Journal of Communication, 37*(3), 269–283.
Harsin, J. (2019). Political Attention: A Genealogy of Reinscriptions. In W. Doyle & C. Roda (Eds.), *Communication in the Era of Attention Scarcity* (pp. 75–113). Palgrave Macmillan.

Hassell, A., & Chin, S. (2023). The Political Influence of Lifestyle Influencers? Examining the Relationship Between Aspirational Social Media Use and Anti-Expert Attitudes and Beliefs. *Social Media + Society*. https://doi.org/10.1177/20563051231211945

Hendricks, V. F., & Vestergaard, M. (2019). *Reality Lost. Markets of Attention, Misinformation and Manipulation*. Springer Nature.

Hight, C. (2022). Deepfakes and Documentary Practice in an Age of Misinformation. *Continuum: Journal of Media and Cultural Studies, 36*(3), 393–410.

Holmstrom, M. (2015). The Narrative and Social Media. *Defence Strategic Communications, 1*(1), 118–132.

Hopkin, J., & Rosamond, B. (2017). Post-truth Politics, Bullshit and Bad Ideas: 'Deficit Fetishism' in the UK. *New Political Economy*. https://doi.org/10.1080/13563467.2017.1373757

Horsthemke, K. (2017). '#FactsMustFall'? Education in a Post-truth, Post-truthful World. *Ethics and Education, 12*(3), 273–288.

Just, N., & Latzer, M. (2017). Governance by Algorithms: Reality Construction by Algorithmic Selection in the Internet. *Media, Culture & Society, 39*(2), 238–258.

Kalpokas, I. (2018). *A Political Theory of Post-Truth*. Palgrave Macmillan.

Kalpokas, I. (2020). Post-Truth and the Changing Information Environment. In P. Baines, N. O'Shaughnessy, & N. Snow (Eds.), *The Sage Handbook of Propaganda* (pp. 71–84). Los Angeles and London.

Kavanagh, J., & Rich, M. D. (2018). *Truth Decay: An Initial Exploration of the Diminishing Role of Facts and Analysis in American Public Life*. The RAND Corporation.

Keller, F. B., et al. (2020). Political Astroturfing on Twitter: How to Coordinate a Disinformation Campaign. *Political Communication, 37*, 256–280.

Kim, J. W. (2018). Rumor Has It: The Effects of Virality Metrics in Rumor Believability and Transmission on Twitter. *New Media & Society*. https://doi.org/10.1177/1461444818784945

Lewandowsky, S., Ecker, U. K. H., & Cook, J. (2017). Beyond Misinformation: Understanding and Coping with the 'Post-truth' Era. *Journal of Applied Research in Memory and Cognition, 16*(4), 353–369.

Mazarr, M. J., et al. (2019). *The Emerging Risk of Virtual Societal Warfare: Social Manipulation in a Changing Information Environment*. The RAND Corporation.

Mazzoleni, G. (2017). Changes in Contemporary Communication Ecosystems Ask for a 'New Look' at the Concept of Mediatisation. *Javnost – The Public, 24*(2), 136–145.

McDermott, R. (2019). Psychological Underpinnings of Post-Truth in Political Beliefs. *PS: Political Science & Politics, 52*(2), 218–222.

McIntyre, L. C. (2018). *Post-Truth*. The MIT Press.

Miller, V. (2020). *Understanding Digital Culture* (2nd ed.). Sage.

Munger, K. (2020). All the News that's Fit to Click: The Economics of Clickbait Media. *Political Communication, 37*(3), 376–397.

O'Shaughnessy, N. (2020). From Disinformation to Fake News: Forwards into the Past. In P. Baines, N. O'Shaughnessy, & N. Snow (Eds.), *The Sage Handbook of Propaganda* (pp. 55–70). Los Angeles and London.

Oleksy, T., et al. (2021). Content Matters. Different Predictors and Social Consequences of General and Government-Related Conspiracy Theories on COVID-19. *Personality and Individual Differences, 168*, 1–7.

Park, C. S., & Kaye, B. K. (2020). What's This? Incidental Exposure to News on Social Media, News-Finds-Me perception, News Efficacy, and News Consumption. *Mass Communication and Society, 23*(2), 157–180.

Pfister, D. S., & Godana, G. D. (2012). Deliberation Technology. *Journal of Public Deliberation, 8*(1), 1–6.

Pierre, J. M. (2020). Mistrust and Misinformation: A Two-Component, Socio-Epistemic Model of Belief in Conspiracy Theories. *Journal of Social and Political Psychology, 8*(2), 617–641.

Sands, S., et al. (2022). Unreal Influence: Leveraging AI in Influencer Marketing. *European Journal of Marketing, 56*(6), 1721–1747.

Schindler, S. (2020). The Task of Critique in Times of Post-Truth Politics. *Review of International Studies, 46*(3), 376–394.

Schoch, D., et al. (2022). Coordination Patterns Reveal Online Political Astroturfing Across the World. *Scientific Reports, 12*, 1–10.

Schwaiger, L., et al. (2022). Mindsets of Conspiracy: A Typology of Affinities Towards Conspiracy Myths in Digital Environments. *Convergence: The International Journal of Research into New Media Technologies, 28*(4), 1007–1029.

Seeliger, M., & Sevignani, S. (2022). A New Structural Transformation of the Public Sphere? An Introduction. *Theory, Culture & Society, 39*(4), 3–16.

Silva, B. S., Vegetti, F., & Littvay, L. (2017). The Elite Is up to Something: Exploring the Relation Between Populism and Belief in Conspiracy Theories. *Swiss Political Science Review, 23*(4), 423–443.

Sloman, S., & Fernbach, P. (2017). *The Knowledge Illusion: Why We Never Think Alone*. Riverhead Books.

Song, H. G., de Zúñiga, H., & Boomgarden, H. G. (2020). Social Media News Use and Political Cynicism: Differential Pathways Through 'News Finds Me' Perception. *Mass Communication and Society, 23*(1), 47–70.

Sunstein, C. R. (2018). *#Republic*. Princeton University Press.

Thorson, K., et al. (2021). Algorithmic Inference, Political Interest, and Exposure to News and Politics on Facebook. *Information, Communication & Society, 24*(2), 183–200.

Toff, B., & Nielsen, R. K. (2022). How News Feels: Anticipated Anxiety as a Factor in News Avoidance and a Barrier to Political Engagement. *Political Communication, 39*(6), 697–714.

Uscinski, J. E., & Enders, A. M. (2022). What Is a Conspiracy Theory and Why Does It Matter? *Critical Review: A Journal of Politics and Society*. https://doi.org/10.1080/08913811.2022.2115668

Vaccari, C., & Chadwick, A. (2020). Deepfakes and Disinformation: Exploring the Impact of Synthetic Political Video on Deception, Uncertainty, and Trust in News. *Social Media + Society*. https://doi.org/10.1177/2056305120903408.

Vaidhyanathan, S. (2018). *Anti-Social Media: How Facebook Disconnects Us and Undermines Democracy*. Oxford University Press.

Valaskivi, K. (2022). Circulation of Conspiracy Theories in the Attention Factory. *Public Communication, 20*(3), 162–177.

van Prooijen, J.-W. (2018). *The Psychology of Conspiracy Theories*. Routledge.

van Prooijen, J.-W., & Douglas, K. M. (2018). Belief in Conspiracy Theories: Basic Principles of an Emerging Research Domain. *European Journal of Social Psychology, 48*(7), 897–908.

Vidigal, R., & Jerit, J. (2022). Issue Importance and the Correction of Misinformation. *Political Communication, 39*(6), 715–736.

Wanless, A., & Berk, M. (2020). The Audience Is the Amplifier: Participatory Propaganda. In P. Baines, N. O'Shaughnessy, & N. Snow (Eds.), *The Sage Handbook of Propaganda* (pp. 85–104). Los Angeles and London.

Webster, F. (2006). *Theories of the Information Society* (3rd ed.). Routledge.

Whittaker, L., et al. (2020). 'All Around Me Are Synthetic Faces': The Mad World of AI-Generated Media. *IT Professional, 22*(5), 90–99.

Wood, M. J., & Douglas, K. M. (2018). Conspiracy Theory Psychology. In J. E. Uscinski (Ed.), *Conspiracy Theories and the People who Believe Them* (pp. 245–256). Oxford University Press.

Woolley, S. (2020). *The Reality Game: How the Next Wave of Technology Will Break the Truth and What We Can Do About It*. Endeavour.

Woolley, S., & Joseff, K. (2020). *Demand for Deceit: How the Way We Think Drives Disinformation*. National Endowment for Democracy and International Forum for Democratic Studies. Available at: https://www.ned.org/wp-content/uploads/2020/01/Demand-for-Deceit.pdf

Zerback, T., & Töpfl, F. (2022). Forged Examples as Disinformation: The Biasing Effects of Political Astroturfing Comments on Public Opinion Perceptions and How to Prevent Them. *Political Psychology, 43*(3), 399–418.

CHAPTER 4

Living with or Through Technology?

Abstract When assessing the impact of technology in a panoramic fashion, the progressivist and nostalgic narratives clearly manifest themselves. On the progressivist side, artificial intelligence (AI) is seen as solving humanity's problems, potentially even ushering in a post-work society in which wealth and comfort can be taken for granted. As a precondition, AI is understood as outperforming humans, making it suboptimal to hold on to human-focused employment and causing a reinvention of the economy itself or supplementing humans with technology. On the nostalgic side, meanwhile, emphasis is on surveillance and data-driven control as leading to the loss of human agency and autonomy. Such datafication processes are seen as merely entrenching existing patterns of power and discrimination and leading to subjection of humans for the profit of technology companies.

Keywords Platform economy • Accelerationism • Surveillance capitalism • Data colonialism

The general sentiment transpires to be one of continuous expansion and pervasiveness of digital technologies and infrastructures, potentially moving towards a world where there will simply be no outside of digital

I. Kalpokas, *Technological Governance and Escapism in Times of Accelerated Change*, Information Technology and Global Governance, https://doi.org/10.1007/978-3-031-60890-2_4

platforms as individuals will either have no alternative but to carry out our activities through them or be constantly surveyed by them even when not using (Susskind, 2018: 153), including through environmental surveillance and surveillance of other individuals. In a similar fashion, van Dijck, Poell and de Waal (2018: 2) postulate 'a connective world where platforms have penetrated the heart of societies—affecting institutions, economic transactions, and social and cultural practices' and, in that process, are driving a change in 'legal and democratic structures' and increasingly *producing* the social world instead of merely reflecting it. After all, the data-richness of today's world is such that 'we daily swim, or perhaps tread-water, in a deep, dark, and forbidding ocean of data that we cannot possibly begin to comprehend' (Herian, 2021: 3). Simultaneously, though, even individual data cannot be seen as purely personal but, instead, 'an amalgamation of previous data processes' (Beer, 2022: 2). In that respect, data is layered upon data as behaviours and interactions are datafied and subsequently fed back by way of the digital environment being imperceptibly structured, thereby shaping further actions that are also subsequently datafied etc. Hence, even the most personal of actions (and the data derived from them) cannot be seen as independent and, therefore, truly personal but, instead, are platform-mediated. While there is little doubt about the technological transformations as such nor is their importance seriously questioned, the interpretation of the aforementioned processes and the value judgements attached to them do differ significantly. In effect, two diametrically opposed (although in some cases also internally diverse) coping strategies are evident: techno-progressivist embrace of AI and associated technologies and techno-pessimist focus on digital harms.

4.1 The Promise of a Better Society to Come

While there is little doubt that the development of digital technologies can bring at least some benefits, techno-progressivist tend to significantly emphasise technological benefits while largely ignoring the threats. In particular, emphasis is typically placed on the emergence of more efficient, affluent, and egalitarian societies in the future or sometimes already at present. For example, when discussing AI-powered assistants, Lind and Dickel (2023: 2) claim that human-machine interactions in such context lead to greater egalitarianism among humans: greater human equality is produced, at least discursively, when existing inequalities are externalised onto such smart assistants. Here the fantasy of mastery and control is clearly evident—not only occluding actual differences between humans, not least in terms of power and wealth, but also seemingly re-restablishing human supremacy as such. Likewise, digital technologies are framed as creating value for societies not only in terms of accelerating economic growth but also by enabling and creating more efficiency in public institutions—essentially, the more public functions are digitised, the more value is expected (Criado & Gil-Garcia, 2019).

On the progressivist side one has the promise of not just more leisure in the context of automation but also improved creative and independently productive capacities, opportunities to do things for their own sake as opposed to toiling for subsistence (Toews, 2021). Instead of substitution one would here encounter ever-shifting entanglements between humans and algorithms that, instead of being oppressive, are seen as complementary in situation-contingent ways (Crosset & Dupont, 2022). A no less ambitious promise would be one of happiness, community involvement, and transcendental pursuits in the post-work world, in which humans will once again have time to engage in noble activities (Etzioni, 2017)—albeit with the ethically dubious caveat of automation having replaced slaves (particularly given Etzioni's direct reference to 'New Athens'). Crucially, AI-generated wealth is foreseen to be such that 'society should be able to flourish like never before', bringing everyone the satisfaction commensurable with them having their dream job but without actually having to work (Tegmark, 2017: 129).

Indeed, for technology optimists a quasi-moral imperative can be devised in terms of something akin to a duty to uphold technological progress. This includes mutual augmentation of humans and machines through synergies with AI in order to create superior hybrid intelligence (Jarrahi et al., 2022). In this way, the expectation would be for scientific advances to go beyond human reasoning, development of solutions to intellectual and moral quagmires, and the ultimate merger of humans and machines, either metaphorically or literally, which can be read, pseudo-religiously, as a path to salvation through technology (Giuliano, 2020: 1015–1016). Indeed, supplementation of humans with technology is often seen as both desirable and unavoidable, leading humans to transcend the fleshy body for the sake of the pure materiality of algorithmically enhanced reason (Herbrechter et al., 2022: 11). Unsurprisingly, the growing importance of technology is also driving calls for doing away with the separation of the human and the machine and the combination of both instead—or at least for viewing both as combinable (Fuchs & Chandler, 2019: 12). One type of such combination could be '[h]uman-non-human assemblages of sensors' that 'enable new forms of responsivity' as well as 'the capacities to see through the breaking down of processes via the development of "correlational machines"' (Chandler, 2019: 34).

Notably, the vision of constant improvement transpires to be self-legitimating as those sceptical of the rosy promises of AI-centric future and the supposed limitless benefits that AI will bring being stigmatised as Luddites or not expert enough, with criticism being presented as any criticism as being directed against improvement of the human condition (Russell, 2020: 153–157). Indeed, it transpires, the constant collection of and acting on data does provide a sense of control and mastery (Herian, 2021: 75), at least under some interpretations. After all, in a world that is changing rapidly and unpredictably, the very idea of data-based prediction and structuration can appear as soothing; likewise, human mastery through the use of, or merger with, technology helps retain fantasies of control. Moreover, such processes of datafied—algorithmic—governance are often 'sold' under the promises of hyper-rationalisation and greater efficiency, which all hark back to and reaffirm the discourse of anthropocentric mastery. The problem, of course, is the biases to which machine learning systems are prone, particularly as they amplify discriminatory trends already present in the data (Shane, 2020: 197). Here, the visions of what is to be done again differ radically: while for the progressivists these are mere teething problems that would be overcome with more data and better algorithms, the nostalgic alternative would, instead, treat such problems as unavoidable and impossible to overcome, with some form of scaling back AI's role in today's societies as a solution, albeit unavoidably a partial one.

More broadly, [illegible] of AI, in the eyes of progressivists, 'is that it will be able to [illegible] accurately, more cheaply and more quickly than humans do them today (Toews, 2021). McAffee and Brynjolfsson (2017: 42), for example, push forward a human substitution narrative with a progressivist spin: for them, it is taken as a fact that once an AI model can be developed and trained, its performance would exceed that of humans working in the same field, with a clear corollary that failure to replace humans under such circumstances would not be justifiable. Almost identically, for Toews (2021), once an activity is successfully and reliably automated and AI can do things better, 'it would be affirmatively harmful and wasteful to have a human in the loop'. In fact, this exceeding of humans is claimed to include not only operational but also foresight performance and problem-solving behaviour: after undergoing the appropriate machine learning processes, according to this line of thinking, 'computers can and will come up with novel solutions that never would have occurred to us' (McAffee & Brynjolfsson, 2017: 116). Hence, getting in the way of technological development can easily be framed as harmful or, at least, dangerous. Crucially, not only the same can be applied to the traditionally productive domains, but also, for example, art is seen as prone to being optimised, with ample opportunities to increase output (Zhou & Lee, 2024). Hence, the efficiency and productivity argument appears to be largely domain-universal.

Digital technologies, particularly platforms and AI-powered tools, are presented as creating a new agile and flexible economy in which novel solutions can be found and implemented without the traditional cumbersome business structures while also enabling near-optimal asset utilisation (Schwab, 2017: 19–20). In this way, it is claimed, '[t]echnology enables greater efficiency, which most people want' (Schwab, 2017: 49). Starting from this premise, it comes as no surprise that the availability and free flow of data are seen as crucial priorities. Effectively, it is often claimed, if only there was more data and it was freed from the corporate silos of today, a global exponential growth in innovation would be unleashed benefitting the entirety of the global society (see e.g. Mayer-Schönberger & Ramge, 2022). In a rather deterministic fashion, it is typically assumed that AI is bound to lead to truly novel discoveries and solutions beyond those that could have been reached by humans; likewise, humans are expected to be exceeded in their capacity in all major domains, from the economy to health to warfare (Kissinger et al., 2021). Overall, technological development is posited as unidirectional, thereby imp[illegible] focus should be on managing the transition, which, if done well, [illegible] a more productive economy and greater wealth for all (se[illegible] & Unger, 2023; Gmyrek et al., 2023; Leliaert, 2023; Georgreva, 2024) rather than on critique, resistance, or attempting a pause.

Of course, value-based questions are not absent for techno-progressivists as at least some of them are willing to admit that alignment with human values is key if AI is not only to exceed humans but also to deliver positive outcomes (Henning, 2021). Also in politics, AI is seen as having an emancipatory and inclusive potential, making sure that all opinions are gathered and accounting for when decisions are being made, thus fostering the hope of reinvigorating democracy (see e.g. Landemore, 2023). In this way, emancipatory capacities are taken to represent the fundamental value structure of AI. Moreover, the transborder nature of AI is sometimes taken to represent a capacity for fostering international cooperation for global good (Gopinath, 2023), thereby seemingly sowing seeds for international cooperation.

Meanwhile, perhaps the most all-encompassing techno-progressivist perspectives are offered by various branches of accelerationism on both political right and left. In terms of the latter, proposals for 'cybernetic socialism' have been floated around at least since the 1970s as a combination of centralised planning with the rationalisation and efficiency offered by computer technology (see e.g. Gardiner, 2022). In more contemporary visions, such left-wing accelerationism is manifested in visions of AI and other forms of automation that are supposed to bring about some form of 'luxury communism' (see, notably, Bastani, 2018; see also Srnicek & Williams, 2015) or some other form of human flourishing. In fact, such visions are not dissimilar from the coping strategies of a century before, such as communist slogans of electrification and industrialisation. Insofar as historical similarities are concerned, there are also clear similarities between accelerationism and the futurist movement of the early twentieth century, with rapid change leading to techno-progressivism with the emphasis on speed and rapid improvement (Noys, 2022: 571–272). Unsurprisingly, then, on the right wing/libertarian side, the development of AI without interference by authorities and guardrails set by developers is seen as in the primary interest of humanity so that new and emergent can be unlocked as well as for progress to unfold as quickly as possible, which is seen as a value in itself (see e.g. Roose, 2023). It is not uncommon for accelerationism to also be accompanied by a drive towards human augmentation with technology and even merger between the two (Noys, 2022: 572). In some more radical versions, accelerationists can even be open to the complete replacement of humans with AI as merely another evolutionary step (Roose, 2023).

Under the progressivist view, then, fears of technology and underlying datafication are irrational: if technological advance equals progress and such advance needs data, then data must be collected or otherwise the economy-growing and life-saving benefits of technology would be jeopardised (Wood, 2019). In this way, one encounters something akin to a moral duty to share data. Otherwise, though, the reasoning could be significantly more prosaic: due to the all-pervasiveness of technology, 'frankly, you *did* passively assent to the digital ecosystem', with datafication being its necessary constituent part (Wood, 2019). When properly contextualised and safeguarded, datafication of population is seen as able to challenge existing exploitative practices of a cultural, societal, and economic nature, thereby allowing those in previously unprivileged positions or otherwise underserved to access services tailored to their needs (Singh, 2023). Even beyond the high-stakes, the effects of datafication and data analytics are seen to add quality to daily life, particularly in terms of helping 'realize desires and aspirations before they occur to us—desires we never even knew we had', leading to another key promise—that of empowerment, getting what we want before or even without knowing we wanted it (Debrabander, 2020: 37) or, at the very least, making consumption quicker and easier.

4.2 Subjection Through Data and Algorithms?

While, as shown above, techno-progressivists embrace ever more intensive datafication, the outlook is markedly different on the nostalgic side. A key sticking point here is enforced transparency through surveillance—as stated by Caplan, and boyd (2018: 4), 'in a world where surveillance is the norm, merely existing in the world means you are structured into the technologies and systems that structure most of social life today'. Surveillance becomes 'environmental'—namely, waved into the fabric of everyday life; the latter also directly implies the need for automation: as vast amounts of data are collected in (nearly) real time, they can only be analyse and corresponding decisions taken if automated tools are put in place (Andrejevic, 2017). According to Beer (2022: 2), this signals the emergence of a 'recursive society'—one in which 'data gathering is so integrated that there is no point of origin'; instead, 'everything is already implicated by the loop upon loop of data processing' so that any data collected and analysed is already based on processes happening on the back of previous data captures and analyses and so on. It transpires that users have already largely accepted—'indeed, domesticated and naturalized'—surveillance and data collection, not because of their enjoyment of being digitally exploited but simply because it has become impossible to imagine alternatives (Kazansky & Milan, 2021: 364), which is seen as an important determinant of human loss of agency. Likewise, one can observe a move from data created by humans to data created by machines, particularly with the advent of the Internet of Things (Nord et al., 2019: 98), thus further deepening the immersion of users in pools of data. Humans, non-human objects and the physical surroundings all interacting in a datafied manner and contributing to a common pool of data that, in turn, enable strategic automated action directed at any combination of those three (Krotov, 2017). In this way, humans are often seen as essentially trapped in data collection and analytics networks.

The knowledge thereby derived is also not without questioning as the relationship between data and the corresponding human person might not be completely straightforward. In particular, the fragmentation/wholeness dichotomy tends to be evoked. For example, it is claimed that 'the human is substituted as an assemblage of their data points which are, in turn, taken as a sign of the real' (Adams, 2021: 177). Hence, the underlying impetus remains geared towards the nostalgic idea of the fullness and wholeness of the human that has supposedly been eroded by technology. Even more explicitly, others would deride 'dehumanisation' through technology and data, i.e. denial or obstruction of 'characteristics that are either uniquely human or constitutive of human nature' (Zawieska, 2020: 277). Instead of human values, the argument goes, the present technological ecosystem of the media would function merely for the purpose of sustaining the dominant ideal of efficiency and profitability on the back of data and robotics (Zawieska, 2020: 875). Such 'dehumanisation' can also manifest itself at individual or group levels when individuals are, paradoxically, *insufficiently* included in the algorithmic ecosystem and are not recognised or misrecognised by it, thereby adversely affecting identities or even closing opportunities, particularly when such lack of recognition takes shape along racial, gendered, or other lines, typically used to separate the privileged and the underprivileged (Jacobsen, 2023).

The question is also one of the value structures embedded in the design of this technologised world, with the more critical voices claiming that instead of some form of universal good, the real underlying model is premised upon the Western-centric legacy of the utility-maximising homo-economicus (see e.g. Elwood, 2021). Indeed, any universal standard achieved, particularly from today's positions of power, would then be seen as merely reproducing and further entrenching 'the classificatory meaning of racialized and gendered intelligence' (Parisi, 2021: 35), as encoded in the models of *artificial* intelligence. Even from a historical perspective, it must be noted, one must always be mindful of 'whose conception of intelligence is modelled with technology or how the idea has been put to work in dividing people between the desired and the undesirable' (Adams, 2021: 176–177). Hence, as also stressed previously in this book, the nostalgic outlook manifests a propensity to posit the anthropocentric model of the human as its object of longing. In this way, while many of the concerns raised are definitely legitimate, the accuracy ethical justifiability of the posited ideal remain problematic indeed.

Likewise, in terms of the structuration and governance of everyday life, algorithmic regulation is presumed to be an emerging order of the day as '*thousands upon thousands* of decisions [...] will be taken every day, decided automatically and executed seamlessly with no right of appeal' (Susskind, 2018: 191). For some, however, such developments even raise the prospect of a dystopian 'world that's devoid of free will' (Frischmann & Selinger, 2018: 209). No less importantly, part of the problem is seen to be within the domain of *capacity* for knowledge: the very 'epistemic possibilities' of decision-making would be changed, with humans now having to rely on AI-generated knowledge without fully understanding either the content of such knowledge or the processes of its generation (Fazi, 2020). Likewise, contrary to the rule of law framework, which implies transparent and identical rules applying equally to all people within a community, algorithmic governance can be seen as capricious and unpredictable but also opaque. It is based on data analytics processes 'in order to identify patterns and correlations that can be used to predict trends and outcomes, and subsequently to perform and automated action' (Elkin-Korel, 2020: 5). Again, the sense of loss of autonomy and independent decision-making capacity is evident.

Indeed, the opacity of regulation through algorithms is a major problem as the code that becomes the enabler of digital architectures (and, through them, of increasingly digital-first lives) is not made available to the public for reasons of commercial secrecy; however, even should we have access to the algorithms and sufficient knowledge to decode them, an additional confounding factor is that at any given time we are at an intersection of multiple effects arising from multiple algorithms simultaneously—and those algorithmic interactions can cause unpredictable results (Dogruel et al., 2022: 1313). Moreover, algorithmic governance can be seen as, in a way, impure, conflating 'the commercial interests of platforms and the governmental functions they perform' (Elkin-Korel, 2020: 6). With the rise in algorithmic structuration of everyday life, it might be excessively difficult not to fall for pseudo-objective micro-instructions provided by algorithm whereby '[c]ode becomes the sole bearer of value' (Ma, 2020: 4). While it would be theoretically possible for 'code as a legal language' to 'resolve legal indeterminacy, producing incontestable truth', in practice this would only mean greater opacity and greater susceptibility to hidden biases (Ma, 2020: 9).

It is not uncommon to encounter claims that in light of ever-expanding delegation and deferral of authority to automated systems, 'common sense becomes embedded in the technological devices' (Frischmann & Selinger, 2018: 203) while datafication, in the shape of the automatic collection, analysis, and use of data, is now seen to be at the heart of the social—indeed, the basis of the latter's construction (Hepp et al., 2018: 5–6; Hepp, 2020: 82). Under this line of argument, algorithms are significantly more than just technological artefacts—in the words of Burke (2019: 12), algorithms are not mere digital agents but 'sociotechnical assemblages' (for an almost identical interpretation, see Elish, & boyd, 2018: 58). In other words, both the way in which algorithms are framed and used and the effects and implications of such use are significantly more than merely technological but also encompass an inter-human dimension as well, arising from specific views and values of social relations and interactions and shaping them back in return. Of course, such a realisation can be added to the list of the previously taken for granted assumptions that are now being challenged: after all, it has traditionally been the case—or, at least, the anthropocentric expectation—that humans would be capable of independently determine their interactions and modes of societal organisation. Following from the latter point, there seemingly transpires to be a great power lurking in the shadows with algorithms having the capacity of shaping the experiences of individuals in a way that is sufficiently imperceptible to escape audience awareness much (if not most) of the time (Park, 2019: 4). Particularly with online platforms that enable the organisational and communicational processes within societies, potential threats encompass those to 'privacy or freedom of expression, manipulation, social discrimination, and other biases' (Dogruel et al., 2022: 1313). Unsurprisingly, then, resistance against the power of algorithms becomes a major theme (see e.g. Velkova & Kaun, 2019: 2), yet again underscoring the nostalgic narrative of the need to take back control and somehow return to more authentic, pre-algorithmic, times.

In the same vein, for Andrejevic (2020: 13, 30), one should, rather straightforwardly, talk of 'automating the social' whereby social processes are replaced with machinic ones. Such transformations, nevertheless, are seen to carry threats of their own, such as 'the displacement of comprehension by correlation, of explanation by prediction and pre-emption, the triumph of efficiency over other social values, and the imperative of total information collection' (Andrejevic, 2020: 30). While some would still insist that meaningful impact of such technological changes is only felt through assemblages of humans and algorithms and, therefore, human augmentation (see e.g. Crosset & Dupont, 2022), others opt for the human substitution scenario, claiming that in light of such developments, even the capacity for meaningful choice seems to be undermined (see e.g. Mhlambi & Tiribelli, 2023). As stressed by Couldry and Mejias (2019: xii), the present time is characterised by for-profit subjection of humans to technology, which would not have been possible 'without radical changes [...] in communication infrastructures, specifically the embedding of computer systems in human life at many levels', whereby technologies have been 'woven into what people do, where they find meaning'. Crucially, it is claimed, as an ever-increasing amount of decisions is being made autonomously—without human participation and even awareness of such decision-making—human autonomy further shrinks, leading to passivity and detachment even in seemingly simple day-to-day tasks that are increasingly delegated to Internet-of-Things devices (van Deursen et al., 2021; see also Dahlgren et al., 2021). In this context, even '[o]ur homes become more open as a node plugged into a global network of information and communication', dominated by, and subjected to the interest of, technology companies (Sadowski et al., 2021: 4). In fact, the home is thereby seen as being transformed by the technological 'non-inhabitants' that undergird daily life and chores (Sadowski et al., 2021: 7). Likewise, through an ever-growing use of AI in governance and decision-making, humans are seen to become passive recipients of instructions and commands, even though such tools are often neither unfair nor unbiased (Janssen et al., 2022; Harrison and Luna-Reyes, 2022).

Unsurprisingly, the effects on human agency are perceived to be such that 'the future is no longer a range of possibilities, but a logically necessary sequence of states of the world' (Berardi, 2021: 42). With the future having become settled, focus is seen to turn to pre-emption, that is, identification of undesirable future outcomes and moving in to alter the predicted course of events (Andrejevic, 2017; Crosset & Dupont, 2022). There is also a non-insignificant loss of dignity implied as a result of 'algorithmic conversion of audiences into objects to mine' (Velkova & Kaun, 2019: 6). In a very similar language, for Zuboff (2019: 10), technology companies, through their products and services, 'lure users into their extractive operations in which our personal experiences are scraped and packaged as the means to others' ends': individuals and their groups thus become raw resources at the epicentre of extractive operations. The extraction-exploitation nexus is taken even further by Couldry and Mejias (2019: x), whose emphasis is on 'the systematic attempt to turn all human lives and relations into inputs for the generation of profit', whereby the totality of human life and experience 'is becoming the target of profitable extraction', which is, in turn, taken to imply '*colonization by data*'. Attention is thereby drawn to 'the continuous extraction of economic value from human life through data (Couldry & Mejias, 2023: 787).

Even with the advent of the supposedly decentralised Web3, the creation of communities and virtual instruments of (co-)ownership primarily are often seen as merely a means to facilitate labour and data extraction (Sadowski & Beegle, 2023). The position criticised by authors roughly subscribing to the above is of data simply being 'out there', open for the taking—reminiscent of the colonial idea of *terra nullius*; in this way, not digitisation *per se* but the conversion of lives to data with the specific purpose of them being mined that constitutes subjection (Couldry and Mejias 2023: 790). Even in presented in a more apolitical and value neutral fashion, 'smart' technologies are still seen as geared towards maximum extraction of data as a resource and optimisation of the commercial logics of profit and speed (Elwood, 2021; see also Zawieska, 2020). In this way, the human as the measure of all things is removed from its supposed position of mastery.

Notably, Andrejevic (2020: 32) sees the prevalence of data-based algorithmic governance as the end of the Enlightenment promise of knowledge becoming rational, objective and verifiable and no longer dependent on belief-based proclamations; for him, the rise of data-rich technology companies represents a step backwards towards 'proclamations by authorities that remain, in terms of understanding, obscure and inaccessible' but, nevertheless, have to be accepted at face value. Likewise, then, the question of power is largely recast since power is now seen to lie with data-based models and automated tools that structure access to reality as well as the actors in possession of them. Rational self-determination, characteristic of anthropocentric exceptionalism, is yet again seen as undermined, with determination by algorithms seemingly becoming the norm. In a sense, therefore, algorithms are interpreted as performing 'like political and cultural institutions' (Velkova & Kaun, 2019: 13). Even in the cultural domain, while platforms 'have made it easier than ever before to access large volumes of cultural goods' it is also the case that 'they have made great strides in personalising the online experience of consuming culture' (Webster, 2019: 1), thereby not only fragmenting cultural experiences but also rendering the setting of cultural values and conventions, in important ways, 'in-human'. Closely related to the above is the argument made by Just and Latzer (2017: 254), whereby reality is inextricably tied to 'the dominance of private governance of global companies' through the large-scale processes of platformisation and datafication. Very similarly, Valaskivi (2022: 168) asserts that 'the logic of the media environment, technological affordances and the ways in which people act in the environment determine what becomes social reality and what does not'. Likewise, for Beer (2018: 476), '[w]e have seen the rise of what might be thought of as a new set of *data intermediaries*, an assemblage of human actors, code, software and algorithms' that ultimately act as gatekeepers for what *is*. In all case, a sense of losing control is evident, with a vision of past human mastery (as opposed to the algorithm) having been the nucleus of nostalgic sentiment.

These new-found sources of power are seen to mostly belong to technology companies that, the narrative goes, ultimately become able not only to foresee but also to model the future based on their data and analytics prowess to the extent that 'the goal now is to *automate us*' (Zuboff, 2019: 8; see also Darmody & Zwick, 2020: 9–10). Similarly, for Gran, Booth and Bucher (2020: 1–2), 'algorithms have become imperative to the internet's infrastructure', capable of producing the choices subsequently made by the humans exposed to algorithmic operations, even to the extent that the very conditions for meaningful participation in public life are actively shaped by algorithms. This shaping, however, tends not to be seen as independent or random but, instead, is specifically directed towards increasing the profits of technology companies (see e.g. Berardi, 2021: 37). In an ultra-competitive technology sphere, it is claimed, companies have little other option than to constantly aim at outdoing each other in terms of locking users in and subsequently subjecting them to corporate interests (Zuboff, 2019: 8). In this vein, automated control is claimed to have become imperceptible or even conflated with a false sense of empowerment, whereby '"good" and autonomous consumer decisions are decisions designed by computational marketing analytic systems, yet experienced as autonomous and self-directed' (Darmody & Zwick, 2020: 10) so that any contingency is eliminated altogether while any notion of independent human action appears to lose its meaning.

Overall, then, it transpires that technology ends up mirroring the impossibility of the rational autonomous Enlightenment self. Instead, the human person emerges as one that is entangled and interdependent with their environment, one whose subjectivity is always contingent and renegotiated. Two equally prominent coping strategies thus emerge: the progressivist one, which embraces such displacement in the name of an even greater projected good (essentially offering an apotheosis of the enhanced human) and the nostalgic one, which laments the passing of supposed human mastery. However, as it also becomes obvious, since both of those strategies provide only a partial account while also ignoring their problematically anthropocentric premises, they still remain simply escapist fantasies.

References

Adams, R. (2021). Can Artificial Intelligence Be Decolonised? *Interdisciplinary Science Reviews, 46*(1–2), 176–197.

Andrejevic, M. (2017). To Preempt a Thief. *International Journal of Communication, 11*, 879–896.

Andrejevic, M. (2020). *Automated Media*. Routledge.

Bastani, A. (2018). *Fully Automated Luxury Communism: A Manifesto*. Verso.

Beer, D. (2018). Envisioning the Power of Data Analytics. *Information, Communication & Society, 21*(3), 465–479.

Beer, D. (2022). The Problem of Researching a Recursive Society: Algorithms, Data Coils and the Looping of the Social. *Big Data & Society*. https://doi.org/10.1177/20539517221104997

Berardi, F. (2021). Simulated Replicants Forever? Big Data, Engendered Determinism, and the End of Prophecy. In N. Lushetich (Ed.), *Big Data – A New Medium?* (pp. 32–45). Routledge.

Brynjolfsson, E., & Unger, G. (2023, December). The Macroeconomics of Artificial Intelligence. *Finance & Development*. https://www.imf.org/en/Publications/fandd/issues/2023/12/Macroeconomics-of-artificial-intelligence-Brynjolfsson-Unger

Burke, A. (2019). Occluded Algorithms. *Big Data & Society*. https://doi.org/10.1177/2053951719858743

Caplan, R., & boyd, d. (2018). Isomorphism through Algorithms: Institutional Dependencies in the Case of Facebook. *Big Data & Society*. https://doi.org/10.1177/2053951718757253

Chandler, D. (2019). Digital Governance in the Anthropocene: The Rise of the Correlational Machine. In D. Chandler & C. Fuchs (Eds.), *Digital Objects, Digital Subjects: Interdisciplinary Perspectives on Capitalism, Labour and Politics in the Age of Big Data* (pp. 23–42). University of Westminster Press.

Couldry, N., & Mejias, U. A. (2019). *The Costs of Connection: How Data Is Colonizing Human Life and Appropriating It for Capitalism*. Stanford University Press.

Couldry, N., & Mejias, A. A. (2023). The Decolonial Turn in Data and Technology Research: What is at Stake and where Is It Heading? *Information, Communication & Society, 26*(4), 786–802.

Criado, J. I., & Gil-Garcia, J. R. (2019). Creating Public Value Through Smart Technologies and Strategies: From Digital Services to Artificial Intelligence and Beyond. *International Journal of Public Sector Management, 32*(5), 438–450.

Crosset, V., & Dupont, B. (2022). Cognitive Assemblages: The Entangled Nature of Algorithmic Content Moderation. *Big Data & Society*. https://doi.org/10.1177/20539517221143361

Dahlgren, K., et al. (2021). Personalization and the Smart Home: Questioning Techno-Hedonist Imaginaries. *Convergence: The Journal of Research into New Media Technologies, 27*(5), 1155–1169.

Darmody, A., & Zwick, D. (2020). Manipulate to Empower: Hyper-Relevance and the Contradictions of Marketing in the Age of Surveillance Capitalism. *Big Data & Society*. https://doi.org/10.1177/2053951720904112

Debrabander, F. (2020). *Life After Privacy: Reclaiming Democracy in a Surveillance Society*. Cambridge University Press.

Dogruel, L., Facciorusso, D., & Stark, B. (2022). 'I'm Still the Master of the Machine'. Internet Users' Awareness of Algorithmic Decision-Making and Their Perception of Its Effect on Their Autonomy. *Information, Communication & Society, 25*(9), 1311–1332.

Elish, M. C., & boyd, d. (2018). Situating Methods in the Magic of Big Data and AI. *Communication Monographs, 85*(1), 57–80.

Elkin-Korel, N. (2020). Contesting Algorithms: Restoring the Public Interest in Content Filtering by Artificial Intelligence. *Big Data & Society*. https://doi.org/10.1177/2053951720932296

Elwood, S. (2021). Digital Geography, Feminist Relationality, Black and Queer Code Studies: Thriving Otherwise. *Progress in Human Geography, 45*(2), 209–228.

Etzioni, A. (2017). Job Collapse on the Road to New Athens. *Challenge, 60*(4), 327–346.

Fazi, M. B. (2020). Beyond Human: Deep Learning, Explainability and Representation. *Theory, Culture & Society*. https://doi.org/10.1177/0263276420966386

Frischmann, B., & Selinger, E. (2018). *Re-Engineering Humanity*. Cambridge University Press.

Fuchs, C., & Chandler, D. (2019). Introduction. In D. Chandler & C. Fuchs (Eds.), *Digital Objects, Digital Subjects: Interdisciplinary Perspectives on Capitalism, Labour and Politics in the Age of Big Data* (pp. 1–20). University of Westminster Press.

Gardiner, M. E. (2022). Automatic for the People? Cybernetics and Left-Accelerationism. *Constellations, 29*, 131–145.

Georgieva, K. (2024, January 14). *AI Will Transform the Global Economy. Let's Make Sure It Benefits Humanity*. IMF Blog. https://www.imf.org/en/Blogs/Articles/2024/01/14/ai-will-transform-the-global-economy-lets-make-sure-it-benefits-humanity

Giuliano, R. M. (2020). Echoes of Myth and Magic in the Language of Artificial Intelligence. *AI & SOCIETY, 35*, 1009–1024.

Gmyrek, P., Berg, J., & Bescond, D. (2023, October). Generative AI and Jobs: Policies to Manage the Transition. *International Labor Organization Research Brief*. https://www.ilo.org/global/research/policy-briefs/WCMS_898026/lang%2D%2Den/index.htm

Gopinath, G. (2023, December). *Harnessing AI for Global Good*. https://www.imf.org/en/Publications/fandd/issues/2023/12/ST-harnessing-AI-for-global-good-Gita-Gopinath

Gran, A.-B., Booth, P., & Bucher, T. (2020). To Be or Not to Be Algorithm Aware: A Question of a New Digital Divide? *Information, Communication & Society*. https://doi.org/10.1080/1369118X.2020.1736124

Harrison, T. M., & Luna-Reyes, L. F. (2022). Cultivating Trustworthy Artificial Intelligence in Digital Government. *Social Science Computer Review, 40*(2), 494–511.

Henning, K. (2021). *Gamechanger AI: How Artificial Intelligence Is Transforming our World*. Springer.

Hepp, A. (2020). *Deep Mediatization*. Routledge.

Hepp, A., Breiter, A., & Hasebrink, U. (2018). Rethinking Transforming Communications: An Introduction. In A. Hepp, A. Breiter, & U. Hasebrink (Eds.), *Communicative Figurations: Transformative Communications in Times of Deep Mediatization* (pp. 3–13). Palgrave Macmillan.

Herbrechter, S., et al. (2022). Critical Posthumanism: An Overview. In S. Herbrechter et al. (Eds.), *Palgrave Handbook of Critical Posthumanism* (pp. 3–26). Palgrave Macmillan.

Herian, R. (2021). *Data: New Trajectories in Law*. Routledge.

Jacobsen, B. N. (2023). Regimes of Recognition on Algorithmic Media. *New Media & Society, 25*(12), 3641–3656.

Janssen, M., et al. (2022). Will Algorithms Blind People? The Effect of Explainable AI and Decision-Makers' Experience on AI-Supported Decision-Making in Government. *Social Science Computer Review, 40*(2), 478–493.

Jarrahi, J. H., Lutz, C., & Newlands, G. (2022). Artificial Intelligence, Human Intelligence and Hybrid Intelligence Based on Mutual Augmentation. *Big Data & Society*. https://doi.org/10.1177/20539517221142824

Just, N., & Latzer, M. (2017). Governance by Algorithms: Reality Construction by Algorithmic Selection in the Internet. *Media, Culture & Society, 39*(2), 238–258.

Kazansky, B., & Milan, S. (2021). 'Bodies not Templates': Contesting Dominant Algorithmic Imaginaries. *New Media & Society, 23*(2), 363–381.

Kissinger, H., Schmidt, E., & Huttenlocher, D. (2021). *The Age of AI and Our Human Future*. John Murray.

Krotov, V. (2017). The Internet of Things and New Business Opportunities. *Business Horizons, 60*, 831–841.

Landemore, H. (2023, December). Fostering More Inclusive Democracy with AI. *Finance & Development*. https://www.imf.org/en/Publications/fandd/issues/2023/12/POV-Fostering-more-inclusive-democracy-with-AI-Landemore

Leliaert, I. (2023, September 20). How to Harness the Power of Generative AI for Better Jobs? Experts Share Their Views. *World Economic Forum*. https://www.weforum.org/agenda/2023/09/how-to-harness-the-power-of-generative-ai-for-better-jobs/

Lind, M., & Dickel, S. (2023). Speaking, but Having No Voice. Negotiating Agency in Advertisements for Intelligent Personal Assistants. *Convergence*. https://doi.org/10.1177/13548565231192100

Ma, M. (2020). The Law's New Language? *Harvard International Law Journal, 61*, 1–9.

Mayer-Schönberger, V., & Ramge, T. (2022). *Access Rules: Freeing Data from Big Tech for a Better Future*. University of California Press.

McAffee, A., & Brynjolfsson, E. (2017). *Machine, Platform, Crowd: Harnessing Our Digital Future*. W. W. Norton & Company.

Mhlambi, S., & Tiribelli, S. (2023). Decolonizing AI Ethics: Relational Autonomy as a Means to Counter AI Harms. *Topoi*. https://doi.org/10.1007/s11245-022-09874-2

Nord, J. H., Koohang, A., & Paliszkiewicz, J. (2019). The Internet of Things: Review and Theoretical Framework. *Expert Systems with Applications, 133*, 97–108.

Noys, B. (2022). Accelerationism: Adventures in Speed. In S. Herbrechter et al. (Eds.), *Palgrave Handbook of Critical Posthumanism* (pp. 557–574). Palgrave Macmillan.

Parisi, L. (2021). Interactive Computation and Artificial Epistemologies. *Theory, Culture & Society, 38*(7-8), 33–53.

Park, C. S. (2019). Does Too Much News on Social Media Discourage News Seeking? Mediating Role of News Efficacy between Perceived News Overload and News Avoidance on Social Media. *Social Media + Society*. doi: https://doi.org/10.1177/2056305119872956.

Roose, K. (2023, December 10). This AI Subculture's Motto: Go, Go, Go. *The New York Times*. https://www.nytimes.com/2023/12/10/technology/ai-acceleration.html

Russell, S. (2020). *Human Compatible: AI and the Problem of Control*. Penguin Books.

Sadowski, J., & Beegle, K. (2023). Expansive and Extractive Networks of Web3. *Big Data & Society*. https://doi.org/10.1177/20539517231159629

Sadowski, J., Strengers, Y., & Kennedy, J. (2021). More Work for Big Mother: Revaluing Care and Control in Smart Homes. *Environment and Planning A: Economy and Space*. doi: https://doi.org/10.1177/0308518X211022366.

Schwab, K. (2017). *The Fourth Industrial Revolution*. Portfolio.

Shane, J. (2020). *You Look Like a Thing and I Love You*. Wildfire.

Singh, R. (2023). The Decolonial Turn Is on the Road to Contingency. *Information, Communication & Society, 26*(4), 803–806.

Srnicek, N., & Williams, A. (2015). *Inventing the Future: Postcapitalism and a World Without Work*. Verso.

Susskind, J. (2018). *Future Politics: Living Together in a World Transformed by Tech*. Oxford University Press.

Tegmark, M. (2017). *Life 3.0: Being Human in the Age of Artificial Intelligence*. Penguin Books.

Toews, R. (2021, February 15). Artificial Intelligence and the End of Work. *Forbes*. https://www.forbes.com/sites/robtoews/2021/02/15/artificial-intelligence-and-the-end-of-work/?sh=2e90907056e3

Valaskivi, K. (2022). Circulation of Conspiracy Theories in the Attention Factory. *Public Communication, 20*(3), 162–177.

van Deursen, A. J. A. M., et al. (2021). Digital Inequalities in the Internet of Things: Differences in Attitudes, Material Access, Skills, and Usage. *Information, Communication & Society, 24*(2), 258–276.

van Dijck, J., Poell, T., & de Waal, M. (2018). *The Platform Society: Public Values in a Connective World*. Oxford University Press.

Velkova, J., & Kaun, A. (2019). Algorithmic Resistance: Media Practices and the Politics of Repair. *Information, Communication & Society*. https://doi.org/10.1080/1369118X.2019.1657162

Webster, J. (2019). Music On-Demand: A Commentary on the Changing Relationship between Music Taste, Consumption and Class in the Streaming Age. *Big Data & Society*. https://doi.org/10.1177/2053951719888770

Wood, M. (2019, August 22). The Ethics of Hiding Your Data from the Machines. *Wired*. https://www.wired.com/story/ethics-hiding-your-data-from-machines/.

Zawieska, K. (2020). Disengagement with Ethics in Robotics as a Tacit Form of Dehumanisation. *AI & SOCIETY, 35*, 869–883.

Zhou, E., & Lee, D. (2024). Generative Artificial Intelligence, Human Creativity, and Art. *PNAS Nexus, 3*, 1–8.

Zuboff, S. (2019). *The Age of Surveillance Capitalism: The Fight for a Human Future at the New Frontier of Power*. Profile Books.

CHAPTER 5

The Less-than-Novel Escapism of the Metaverse

Abstract Visions for a future metaverse present perhaps the clearest manifestation of progressivist escapism. Not only the postulated emergence of an alternative digital world is framed as the extension of the fantasy of human dominance and exploitation of the environment but also it represents a fantasy of unlimited agency and fulfilment by simulating any experience imaginable. Moreover, through virtualisation of everything, from assets to dominance and even slavery, the metaverse opens up a quasi-colonial race towards new frontiers of appropriation. In this sense, despite its alleged novelty, the metaverse merely serves to reinvigorate some of the paradigmatic anthropocentric assumptions and principles that have historically been characteristic of Western societies.

Keywords Virtual worlds • Immersion • Presence • Experience • Simulation

It must be admitted straight away that there is no such thing as a single and unified vision for *the* metaverse—instead, one can find a set of 'partly overlapping, partly compatible ecosystems' and, likewise, partly overlapping approaches and conceptual frameworks (Jacobides et al., 2024: 123; see also Wang et al., 2024). However, it is equally clear that, differences,

I. Kalpokas, *Technological Governance and Escapism in Times of Accelerated Change*, Information Technology and Global Governance, https://doi.org/10.1007/978-3-031-60890-2_5

notwithstanding, an overlap in terms of the underlying ideological thrust is evident, and it is one of overwhelming technological optimism. As such, the metaverse performs a further function of technological mirroring—that of reflecting back an idealised (and, in some ways, radicalised) anthropocentric image. In this way, it functions as an escapist fantasy of the technological sublime. As such, the metaverse can be seen as falling under the category of what Dahlgren et al. (2021): 1161) call a general trend towards imagining a techno-hedonist future of effortless efficiency. Nevertheless, due to its virtual character, the metaverse can be seen as more *literally* escapist than other techno-progressivist visions of the future.

Of course, the metaverse, as a concept, has already undergone different stages of the hype cycle, having soared to popularity around 2021 and early 2022, only to be replaced *the* buzzword by generative AI later that year. As such, it is very easy to postulate the death of the metaverse and question the very reasoning behind any further discussions of the subject. Nevertheless, it is claimed here that there is more to the metaverse than the metaverse itself: part of the reason behind its surge in popularity was not only what it *is* but also what it *represents*. In a sense, visions for the metaverse can be seen to constitute an apotheosis of progressivist technological escapism. Moreover, it must also be noted that while 'the big thing' of the metaverse might have lost its rhetorical power and financial clout, work on virtual environments and immersive spaces continues nevertheless, although perhaps without some of the grandiose claims, characteristic of the venture capital-funded technology world.

5.1 Conceptualising the Metaverse

It is fair to say that an overarching definition of the metaverse—one that would satisfy all of its present and past adherents—is lacking, although most, as Richter and Richter (2023: 9) note, would include an integration of, among other things, 'the latest advances in [...] virtual and augmented reality, haptic feedback, artificial intelligence, and blockchain technology'. Broadly, the metaverse can be classified as belonging to the domain of Extended Reality (XR), which, in turn, comprises of Virtual Reality (VR) and Augmented Reality (AR). The latter might be interpreted as offering a somewhat lower degree of escapism, since it does not block out the user's physical environment completely (Rauschnabel et al., 2019: 44)—instead, it adds a computer-generated overlay in real time in reaction to the user's actions in the physical space (Scholz & Duffy, 2018: 12; Steffen et al., 2019: 687). As such, the user acquires the capacity to manipulate their physical surroundings in an unprecedented way (Goebert & Greenhalgh, 2020). Meanwhile, VR allows for immersion in completely new realities detached from the user's physical surroundings (Tussyadiah et al., 2018: 141), thereby having as its goal the capacity to 'deliver high quality on cognitive stimulation, emotional pleasure and behavioral activation' within content that is relevant and relatable (Tsai, 2020: 1082). Hence, VR is to be understood as an all-encompassing multi-dimensional computer generated environment or a network of virtual worlds (Egliston et al., 2024) which, by definition, works by excluding the individual from everything that is not part of the simulation (Steffen et al., 2019: 689–690). In this way, the VR (although, to an extent, also the AR) element of the metaverse opens up the possibility for escape by offering a vision of easy and unlimited transformation, reinvention, and even new genesis—this time, of a virtual world—which is highly progressivist because of the centrality of technological innovation that is projected to continue on a linear part of development and improvement (Mosco, 2023: 169–170).

Of course, the 'bare bones' of technological affordances are not enough—instead, coded affordances and content are crucial for the imaginary of the metaverse but also, of course, for its ultimate success or lack thereof. Here, Ball (2022) provides perhaps the most extensively reasoned account, which includes massive scale, interoperability, real-time rendering of fully immersive 3D experiences, synchronicity and persistence of such experiences and associated data flows, the ability to accommodate an unlimited number of simultaneous users, a virtual economy that is fully functioning and capable of sustaining itself, and sense of presence on the individual level as the key features of the metaverse. A similar set of features is also put forward by Wang et al. (2024: 6) who identify them as immersiveness (including multisensory 3D experiences), shared nature (persistent and synchronised in real time for all users), interoperability (freedom of movements across worlds), and scalability (capacity to accommodate an essentially unlimited number of users and experiences simultaneously) as key features of the metaverse. Likewise, for Golf-Papez et al. (2022: 3), 'the metaverse is an ecosystem of interconnected, shared digital and physical environments that can be experienced synchronously, persistently, and interoperably, and in which physical and technology-enhanced realities are seamlessly combined'. What all of these sets of features have in common, though, is the capacity for an unlimited number of individuals to share the perception that they have escaped the physical world into a virtual world and to have fulfilling experiences there so that return to the physical world no longer seems attractive.

In light of the above, the metaverse would not be a mere substitute for physical reality but, instead, would constitute a completely *new* reality, the value of which is premised upon users' capacity to enter into synthetic relationships and have synthetic experiences within equally synthetic worlds that are capable of generating 'cognitive, emotional, behavioral, sensorial, and social responses' of the users that are not inferior to those experienced in the physical world (Golf-Papez et al., 2022: 3; see also Ball, 2020). In the same vein, it must be noted that the term 'synthetic' here is not used in a way that carries an evaluative meaning: it merely pertains to the object thus described being computer-generated, as in the broader term 'synthetic media'. Indeed, AI-generated synthetic media can be expected to underpin virtual experiences, for the simple reason that if such virtual world are to be potentially infinite and unrestricted as well as malleable and adjusting to user actions in real time so that no glitches and delays ruin the smoothness and immediacy of experience, content generation processes simply have to be automated and operate at machinic speeds (Hackl, 2023: 33; see also Wang et al., 2024).

Notably, the key concepts describing the attractiveness of the metaverse are immersion and presence. Immersion is perhaps best described as a perception of 'being "in" a real or virtual experience' in a way that 'involves physical and mental participation and implies getting away from everyday experience, playing a different role or taking on a new identity' (Hudson et al., 2019: 461). While there are multiple aspects facilitating immersion, such as 'realism and interactivity of the virtual environment, ease of use of the technological interface, the identification of the player with the environment, enjoyment, and social interactions', users are only truly immersed when they cease to think about the world they are in but, instead, are carried by the 'flow' of what is happening, with a corresponding emotional valence (Hudson et al., 2019: 461). Meanwhile, the feeling of presence can be defined as 'a psychological state in which the virtuality (artificiality) or an experience is unnoticed'—one that takes shape through a complex framework characterised by an 'interplay of multi-sensory information and various cognitive processes' (Tussyadiah et al., 2018: 141–142). At the core here is the illusion of no mediation taking place, with not only the physical surroundings but also the technology disappearing from one's field of perception as if neither was here and the virtual experience constituted all that exists (Tussyadiah et al., 2018: 142). Of course, this loss of awareness of technological intermediation not only is (obviously) misguided but also masks the deeper practices of datafication and surveillance whereby the underlying hardware cannot function in any other way *but* by tracking a user's every move and longitudinal behavioural patterns in order to render the virtual experience in real time and do so in a way that seamlessly corresponds to the user's actions (Riva et al., 2019: 88–89). Of course, true presence necessitates more than simple visual simulation of alternative worlds—instead, haptic technologies (tools to simulate physical sensations) must also enter into the fore, since without haptic richness and multisensory feedback the feeling of presence will unavoidably be diminished (Steffen et al., 2019: 690). In this way, the aim can be seen to revolve around the need to create a rich and all-encompassing environment for virtual escape into a technologically enabled alternative world which, paradoxically, necessitates the users to lose any perception of the same technology.

A crucial part of any virtual world (and key constitutive feature of both immersion and presence) is the presence of rewarding social interactions. Hence, another goal (although one less dependent on the technology companies that can act, at most, as enablers and facilitators but more pertaining to the users themselves) is the creation of a new, (as yet) non-existent social world (Sadowski & Beegle, 2023: 11). If that succeeds and presence becomes co-presence, such virtual environments would truly become constitutive of human lives to the extent that the distinction between the physical and the virtual 'will soon become moot, if still technically accurate' (Rubin, 2020: 124; see also Chan, 2022: 129–130). That, if course, is not completely unprecedented: after all, it must be stressed that technologies are already fundamentally interwoven with everyday activities, habits, movements, and living conditions (Chan, 2022: 175). Hence, what the metaverse promises is a radicalisation of the same trend.

One could expect a significant degree of escape from the confines of the physical environment of the user as greater immersion implies a corresponding reduction in awareness of one's physical surroundings (Han et al., 2022). Indeed, it transpires that users typically do understand virtual environments as existential situations, that is, worlds in which they can engage in purposeful action, and likewise develop a virtual subjectivity (Gualeni & Vela, 2020). In fact, the enmeshment with digital technologies and withdrawal from the physical world can also be reframed as being not an anti-social trend but, instead, a new form of interconnectedness and sociality (Wong, 2020). Even if, at least as of now, questions tend to be asked as to the degree of reality that virtual actions and artefacts have, there is an expectation of a time to come when 'the physical and virtual worlds [will] collide and morph into a broader single entry: *existence*' (Greengard, 2019: 209). Characteristically of escapist fantasies, therefore, the present reality (even the present mode of existence) is deemed to be (no longer) satisfactory, thereby creating a yearning for improvement. And, as is typical of *progressivist* fantasies, such improvement is seen to await in the (already foreseeable) future.

Just like today's video games are by design 'structured around data and feedback' (Tulloch & Johnson, 2022: 925), the metaverse can be expected to be a fully tailored experience, adapted in real time to the data being collected here and now. While this might well play into the fears of surveillance and pervasive datafication outlined previously in this book, the latter are often conveniently overlooked in visions for the metaverse, focusing instead on the escapist tendencies inherent in the end result. Likewise often omitted is the manipulative potential of the metaverse, particularly with regard to domains such as 'advertising, false news, and political propaganda' (Greengard, 2019: 183). For example, even when considering virtual influencers and other automated agents that can be expected to populate the metaverse, while fulfilment of commercial interests would likely be a primary driver for their creation, political influence and the shaping of users' attitudes towards societal issues would also likely play a not insignificant role (Block & Lovegrove, 2021: 269). Hence, the techno-progressivist escape offered by the metaverse as it is typically portrayed can also be seen to mask immediate short-term threats.

More critical voices can be heard as well, particularly emphasising how virtual worlds are sold as a quasi-magical escape from 'a world of increasing wealth inequality, environmental disaster, and political instability' (Gault, 2021; see also Han et al., 2022: 13). Moreover, such escape happens less-than-organically, notably, under the terms set out by the technology companies that, by offering everyone a virtual version of a desirable lifestyle would be potentially able to preclude demands for real change (Gault, 2021). While providing an important counterbalance to the dominant techno-optimism, such views, nevertheless, are also prone to passing into the nostalgic camp by emphasising how VR, brain-machine interfaces and the like would ultimately leave the human body nothing more than a 'meat peripheral' (Gault, 2021)—although, as noted below, the progressivists tend to come to a similar conclusion, just in a different mood.

5.2 What Is in the Metaverse?

Being a synthetic environment, the metaverse is, of course, also a highly malleable one—not only fully immersive but also, no less importantly, one that surpasses the limitations of physical space (Al-Sharafi et al., 2023). Indeed, the metaverse is seen to offer quasi-complete customisation not only of one's environment but also of one's self (through an avatar) and of social interactions that can take place with both humans and AI-powered avatars. The latter, despite their artificiality, would likely play an important role in ensuring that individuals are at the centre of their personal virtual universes. It is worth noting here that virtual agents can still have an impact and be understood as authentic despite having no physical correlate to refer to (Koles et al., 2024). Paradoxically, this is precisely because they are not free-standing self-sufficient entities but, instead, are created in relation to their audiences who participate in co-creation, meaning that a virtual personality 'exists as long as she is fuelled by her audience' (Jiang, 2022: 4). It is this user investment in co-creation that drives perceived value. The same applies to engagement with the digital environment more broadly, whereby having a subjectively relevant stake leads to perception of virtual reality as existential (Gualeni & Vela, 2020: 4). To an extent, such alternative reality can be even seen as empowering (or can be easily framed as such) because individuals can expect to be in full control over their lives—to a larger extent than it would be possible in the offline world (van Rijmenam, 2022). Hence, escapism is here rendered possible by way of creating meaning even in the absence of material form (or, rather, potentially even *because of* such absence).

Indeed, being freed from the constraints of the physical world features prominently as a selling point for virtual experiences. Unsurprisingly, human subjectivity is typically displayed as progressively augmented and freed from nature through technological advancements that collectively allow for digital technological augmentation (Gualeni & Vela, 2020: 71). To reach this stage, the threshold of transition needs to be managed first, for which a downgrade of the present and an upgrade of the future is often an effective strategy. In order to be sellable to the general audience, the metaverse has to be seen as a fundamental improvement that 'disrupts time by inviting users to leave the desultory present, with all the treats to human survival, and invites users to enter a future world where climate change, pandemic, and the spectre of nuclear annihilation do not exist' (Mosco, 2023: 170). Here as well, the transition is often marked by crude mechanistic analogies, whereby the physical embodiment of the individual, including the 'hardware' on which the mind runs, is seen as yet another machine, albeit an inherently limited one, which can be plugged into computers as a means of upgrading so that in the end the human mind would have the capacity to 'access and process information in ways we can't even begin to understand today' (Narula, 2022: ix–x). Though such allegedly inevitable upgrades, one can expect to gain access to 'a future in which our minds are untethered from the limits of our bodies' (Narula, 2022: 218). Here, techno-progressivist escapism can be seen to pass another benchmark: complete technology-enabled escape from the body into the digital realm.

Moreover, as the metaverse removes physical constraints, potential for development and growth is likewise seen as limitless (see e.g. Al-Sharafi et al., 2023), offering 'infinite new possibilities for happiness, expanded intelligence, and intrinsic growth on a scale that we can't even imagine today' (Narula, 2022: 225). In this way, the line between the present and the mediated is further blurred (Klowait, 2019: 617) in what is claimed to be 'a parallel plane of existence' (Ball, 2022: 16). The stated goal here would be unleashing the power of the human mind and imagination in ways that are unconstrained by the laws of the physical world (Hackl, 2023: 149), particularly as the world of atoms would then be replaced with that of bits, both carrying essentially the same amount of 'reality' (Ball, 2022: 67). Under such circumstances, immersion in virtual worlds is projected to replace day-to-day offline activities and social, economic, and political interactions (Terry & Keeney, 2022). As such, then, connecting and identity-building are also expected to acquire a virtual form—one marked with significant user-centricity (Lucia et al., 2023). Indeed, it is a characteristic feature of visions for the metaverse that every user gets the opportunity to be at the centre of their own virtual experience, with the latter open to curation and, it would be fair to say, domination (see, characteristically, Terry & Keeney, 2022: 5).

In light of the above, the metaverse offers the opportunity to free individuals from the constraints imposed on them by material environmental factors and laws of nature, thereby enabling them to be wherever they happen to want to be and do whatever crosses their minds—or at least as long as it is in line with the terms of service of a particular platform (van

Rijmenam, 2022). As alleged by Steffen et al. (2019: 685), 'users adopt virtual and augmented reality because they afford activities that are impossible or advantageous when compared to the activities afforded by the physical reality'. One way to illustrate this would be a situation in which a user is able to use the metaverse to live out experiences and lifestyles that are either unaffordable or physically impossible (or both) offline (see, broadly, Terry & Keeney, 2022: 6; see also Ball, 2022: 254–260), such as having an apartment on top of the Eiffel Tower or living on a coral reef as a mermaid. Moreover, the metaverse can be seen as an escape from issues around poverty and overpopulation, again by merely providing a simple technological fix: as Narula (2022: 225) claims, 'we would literally be able to have trillions upon trillions of people living in the equivalent of luxury inside a simulation' (Narula, 2022: 225). Hence, there seemingly is no need to address offline social problems—all of them can be overcome by simply uploading humans onto a virtual simulation. However, instead of acknowledging this escapist element, the framing here tends to be akin to pursuit of 'peak fulfilment' and emergence of new societies that would not otherwise be humanly possible (Narula, 2022: 226–227).

Mere individual-level gratification would not suffice, however, if it was not for a social element in the metaverse. As rather emphatically stressed by Rubin (2020: 111), '[t]he space you're in won't matter nearly as much as who you're there with'. Crucially, such virtual sociality is expected to be not only rewarding but also persistent, particularly due to VR's ability to create shared memories that feel embodied even though they have nothing to do with the physical world (Rubin, 2020: 138). It is, therefore, unsurprising that interaction with virtual avatars (of both real persons and AI-enabled characters) might end up becoming the dominant mode of interaction. After all, social experiences in VR are foreseen to be capable of inducing 'a stronger emotional response than simply interacting with a game character or another computer-generated construct' (Rubin, 2020: 124). While it might be argued that the benchmark for claiming an improvement is, in this case, not particularly high, such predication of meaningful sociality is one of the key factors behind ideas of the metaverse and other VR experiences. Furthermore, virtual spaces would also be characterised by relative indistinguishability between human and AI avatars, with an increasing role played by such AI personalities as virtual agents (Hackl, 2023: 48). Yet, very tellingly, a key characteristic of such AI avatars is presented as knowing 'how to please' (Hackl, 2023: 58), thereby further reiterating the fantasy of human domination through (and of) technology. Indeed, the very existence of AI avatars would be specifically for the purpose of human enjoyment and consumption (Ragavan, 2021). In therefore comes as no surprise that human—AI avatar interactions, at least given current representations, could also end up including highly morally dubious forms, such as owning virtual slaves, even if the term itself is unsurprisingly avoided in mainstream visions for the metaverse, with authors opting for something less loaded, such as 'utility bots' (Terry & Keeney, 2022: 6–9). In this way, technologised escape is projected in terms of not only the ambit of human interactions but, even more fundamentally, the norms and responsibilities inherent in human interactions. Of course, such tendencies are not completely new either—owning digital slaves might be, in a way, comparable to the current practice of boosting productivity through outsourcing some tasks to micro-workers performing click tasks on service platforms, often in digital sweatshop conditions. However, in the metaverse the (neo-)colonial fantasy of domination and of the entire world being subjected to the achievement of one's goals is maximised. For this reason, even if virtual interaction is key, much of it should not be idealised since it is likely to take place along the lines of Grear's (2015) critique of anthropocentrism outlined earlier in this book: the privileged *anthropos* as little more than a privileged subset of those who are taken to matter.

Again, it is no surprise that much of the functionality of today's virtual reality is premised upon a narrative of efficiency and time pressure: as stressed, in a slightly different context, by Chambers (2021: 1181), 'technology compresses time to generate a sense of time expiring, slipping away and "running out"', subsequently creating a demand for ever new technological solutions for, effectively, buying back time. The promise of work or leisure without limits is also strongly premised upon a disintegration of boundaries in media use: 'a blurring of work and home zones, spurred by notions of temporal excess, absorption, immersion and a squandering of time' (Chambers, 2021: 1182). Such loss of traditional notions of both distance and connection is also noted by Greengard (2019: xvi) for whom, with the advent of VR, participation in activities becomes detached, suspended among 'a group of people doing something together while separated', with every person who partakes in such action being 'isolated in his or her own virtual-reality world'. In absence of clearly determinable boundaries and with the whole of the lived environment becoming an endless quest for greater efficiency, productivity, self-improvement, and other forms of achievement, the instrumentalisation of time, space, and fellow entities as mere resources is an outcome that can be expected, with the metaverse offering an escape of a very specific kind: not an escape from a lifestyle characterised by exploitation of self and others but an escape from the drudgery of it all—the emergence of assort of Disneyland of exploitation.

The promise tends to be one of a liberation of the self *from the self*, particularly in terms of any resemblance between users and their avatars simply becoming a matter of choice. Hence, both bodies and places cease to be limiting factors for experiences, opening ways for novel digital lives (Škola & Liarokapis, 2021) while conventionally real and virtual environmental affordances mix to create completely new experience worlds (Heemsbergen et al., 2021). The self is thereby re-rendered as a project in the making or a continuous journey of progressive self-realisation and self-improvement, primarily though creation of new identities via virtual bodies, that is, avatars (Chan, 2022: 142). Similarly, for Procter (2021: 60), at the heart of virtual existence is self-manifestation, self-actualisation, and self-presentation through the avatar, offering 'nearly limitless possible options for appearance, character, back story, and even species'. Indeed, avatars are typically projected as giving users 'the freedom to be anything they want' (Hackl et al., 2022: 111). Such augmentation and customisation of the self (see also Ball, 2022; Hackl, 2023) can be seen as the ultimate expression the abovementioned technological escape from the body as such by opening the latter to constant reinvention. While there is a clear emancipatory element in such visions—unshackling one's identity from the physical body, particularly in cases when the two are at a disjoint—while also underscoring an element of playfulness and creativity (see e.g. van Rijmenam, 2022; Al-Sharafi et al., 2023), there can also easily be more than an element of conformity with hegemonic standards and even ableism as the avatar is seen as a possibility to overcome underprivileged identities and disabilities and to present oneself as in line with the mainstream (see e.g. Lemley & Volokh, 2018: 1067). Interestingly, though, despite the emphasis on freedom and creativity, it transpires that individuals still tend to create avatars that roughly resemble their own physical selves, albeit in an idealised form (Hadi et al., 2024).

In a similar manner, tangible assets are projected to be replaced with virtual ones (Terry & Keeney, 2022) as virtual fulfilment through virtual participation and virtual consumption is presented to become the new value-ascribing factor (Narula, 2022). Technology would then end up enabling 'an enhanced experience, more interactivity with environments, more connection with communities, and more rewards', thus implying that being stuck in the tangible domain would, in the end, become counter-productive (Hackl et al., 2022: 202). Not only that—if tangible assets and offline wealth are seen as naturally constrained in terms of growth (by e.g. resource and labour availability, productivity etc.), the techno-progressivist view espoused by metaverse advocates puts forward a vision of unlimited growth and development through technology, thereby creating a potentially unlimited amount of virtual wealth for all (Hackl et al., 2022: 206; see Koohang et al., 2023). The metaverse would then become, not without colonialist undertones, 'an infinite blue ocean of opportunity' (van Rijmenam, 2022: 32) or 'a new frontier where exploration is limitless' (Hadi et al., 2024: 159). Yet again, techno-optimist escapism can be seen as enabling the fantasy of domination and unlimited exploitation as a precondition to the projected (virtual) prosperity of the future.

Moreover, there is a further observation that needs to be made: quite tellingly, the focus on virtuality and detachment from the physical world that characterises the metaverse leaves no room for environmental considerations. Not only does the metaverse serve as an easy escape from the climate emergency that we are currently living through—by implying that it can simply be ignored by escaping into virtual worlds—but also the environmental footprint of the metaverse itself tends to be largely ignored. While, for example, Hackl (2023: 34) asserts, without any justification or elaboration, that the metaverse constitutes 'a more environmentally responsible alternative to traditional work models and design flows', substantive engagement with the materiality of the metaverse is hard to come by. Of course, some metaverse advocates are not oblivious to the physical components of the metaverse: for example, as Ball (2022: 202) stresses, 'digital first requires the physical first', particularly in terms of hardware and connectivity infrastructure. However, physical components are still seen to play an ancillary—enabling—role for the *real thing*; that is, the immaterial virtual experience of the metaverse and mitigation of adverse impact is left to the silver bullet of future technological innovations.

Overall, then, the metaverse offers a techno-progressivist escape not only from the immediate concerns that dominate the offline world but also, notably, from the self, or at least from the embodied aspect thereof. In doing so, the mainstream visions for the metaverse also reproduce some of the key anthropocentric assumptions outlined earlier in this book, radicalising them along the way. Typically of a coping strategy, then, the metaverse represents a utopian spin on the 'business as usual'—by picking up on some of the taken for granted assumptions that are otherwise destabilised and constructed a fantasy in which they nevertheless become characteristic features of the future. Likewise, the dominant framing of the metaverse again reveals a mirroring function inherent in technology, this time reflecting the prevalence of domination-focused and exploitative impulses that are seemingly normalised the techno-progressivist project.

References

Al-Sharafi, M. A., et al. (2023). Drivers and Barriers Affecting Metaverse Adoption: A Systematic Review, Theoretical Framework, and Avenues for Further Research. *International Journal of Human-Computer Interaction*. https://doi.org/10.1080/10447318.2023.2260984

Ball, M. (2020, January 13). *Metaverse: Where to Find It, Who Will Build It, and Fortnite*. https://www.matthewball.vc/all/themetaverse

Ball, M. (2022). *The Metaverse and How It Will Revolutionize Everything*. Liveright Publishing Corporation.

Block, E., & Lovegrove, R. (2021). Discordant Storytelling, 'Honest Fakery', Identity Peddling: How Uncanny CGI Characters are Jamming Public Relations and Influencer Practices. *Public Relations Inquiry, 10*(3), 265–293.

Chambers, D. (2021). Emerging Temporalities in the Multiscreen Home. *Media, Culture & Society, 47*(3), 1180–1196.

Chan, M. (2022). *Digital Reality: The Body and Digital Technologies*. Bloomsbury Academic.

Dahlgren, K., et al. (2021). Personalization and the Smart Home: Questioning Techno-Hedonist Imaginaries. *Convergence: The Journal of Research into New Media Technologies, 27*(5), 1155–1169.

Egliston, B., Carter, M., & Clark, K. E. (2024). Who Will Govern the Metaverse? Examining Governance Initiatives for Extended Reality (XR) Technologies. *New Media & Society*. https://doi.org/10.1177/14614448231226172

Gault, M. (2021, February 15). Billionaires See VR as a Way to Avoid Radical Social Change. *Wired*. https://www.wired.com/story/billionaires-use-vr-avoid-social-change

Goebert, C., & Greenhalgh, G. P. (2020). A New Reality: Fan Perceptions of Augmented Reality Readiness in Sport Marketing. *Computers in Human Behavior, 106*, 1–11.

Golf-Papez, M., et al. (2022). Embracing falsity through the Metaverse: The Case of Synthetic Customer Experiences. *Business Horizons*. https://doi.org/10.1016/j.bushor.2022.07.007

Grear, A. (2015). Deconstructing *Anthropos*: A Critical Legal Reflection on 'Anthropocentric' Law and Anthropocene 'Humanity'. *Law and Critique, 26*, 225–249.

Greengard, S. (2019). *Virtual Reality.* The MIT Press.

Gualeni, S., & Vela, D. (2020). *Virtual Existentialism and Subjectivity in Virtual Worlds.* Palgrave Macmillan.

Hackl, C. (2023). *Into the Metaverse: The Essential Guide to the Business Opportunities of the Web3 Era.* Bloomsbury.

Hackl, C., Lueth, D., & Di Bartolo, T. (2022). *Navigating the Metaverse: A Guide to Limitless Possibilities in a Web 3.0 World.* Wiley.

Hadi, R., Mehmad, S., & Park, E. S. (2024). The Metaverse: A New Digital Frontier for Consumer Behavior. *Journal of Consumer Psychology, 34*, 142–166.

Han, D.-I. D., Bergs, Y., & Moorhouse, N. (2022). Virtual Reality Consumer Experience Escapes: Preparing for the Metaverse. *Virtual Reality.* https://doi.org/10.1007/s10055-022-00641-7

Heemsbergen, L., Bowtell, G., & Vincent, J. (2021). Conceptualising Augmented Reality: From Virtual Divides to Mediated Dynamics. *Convergence: The International Journal of Research into New Media Technologies, 27*(3), 830–846.

Hudson, S., et al. (2019). With or Without You? Interaction and Immersion in a Virtual Reality Experience. *Journal of Business Research, 100*, 459–468.

Jacobides, M. G., et al. (2024). Building Synthetic Worlds: Lessons from the Excessive Infatuation and Oversold Disillusionment with the Metaverse. *Industry and Innovation, 31*(1), 105–129.

Jiang, C. (2022). The Making of Popstar Fembots: Participation, Co-Creation, or online Cultural Exploitation? *Hybrid: Journal of Arts and Human Mediations, 8*, 1–12.

Klowait, N. O. (2019). Interactionism in the Age of Ubiquitous Telecommunication. *Information, Communication & Society, 22*(5), 605–621.

Koles, B., et al. (2024). The Authentic Virtual Influencer: Authenticity Manifestations in the Metaverse. *Journal of Business Research, 170*, 1–20.

Koohang, A., et al. (2023). Shaping the Metaverse into Reality: A Holistic Multidisciplinary Understanding of Opportunities, Challenges, and Avenues for Future Investigation. *Journal of Computer Information Systems, 63*(3), 735–765.

Lemley, M. A., & Volokh, E. (2018). Virtual Reality and Augmented Reality. *University of Pennsylvania Law Review, 66*(5), 1051–1138.

Lucia, B., Vetter, M. A., & Adulbofour, I. K. (2023). Behold the Metaverse: Facebook's Meta Imaginary and the Circulation of Elite Discourse. *New Media & Society.* https://doi.org/10.1177/14614448231184249

Mosco, V. (2023). Into the Metaverse: Technical Challenges, Social Problems, Utopian Visions, and Policy Principles. *Javnost – The Public, 30*(2), 161–173.

Narula, H. (2022). *Virtual Society: The Metaverse and the New Frontiers of Human Experience.* Currency.

Procter, L. (2021). I Am/We Are: Exploring the Online Self-Avatar Relationship. *Journal of Communication Inquiry, 45*(1), 45–64.

Ragavan, S. (2021, January 25). Virtual Influencers Blur the Line Between Fantasy and Reality. *PR Week.* https://www.prweek.com/article/1705359/virtual-influencers-blur-line-fantasy-reality?bulletin=asia-bulletin

Rauschnabel, P. A., Felix, R., & Hinsch, C. (2019). Augmented Reality Marketing: How Mobile AR-Apps Can Improve Brands through Inspiration. *Journal of Retailing and Consumer Services, 49*, 43–53.

Richter, S., & Richter, A. (2023). What Is Novel About the Metaverse? *International Journal of Information Management, 73*, 1–11.

Riva, G., Wiederhold, B. K., & Mantovani, F. (2019). Neuroscience of Virtual Reality: From Virtual Exposure to Embodied Medicine. *Cyberpsychology, Behavior, and Social Networking, 22*(1), 82–98.

Rubin, P. (2020). *Future Presence: How Virtual Reality is Changing Human Connection, Intimacy, and the Limits of Ordinary Life.* Harper One.

Sadowski, J., & Beegle, K. (2023). Expansive and Extractive Networks of Web3. *Big Data & Society.* https://doi.org/10.1177/20539517231159629

Scholz, J., & Duffy, K. (2018). We ARe at Home: How Augmented Reality Reshapes Mobile Marketing and Consumer-Brand Relationships. *Journal of Retailing and Consumer Services, 44*, 11–23.

Škola, F., & Liarokapis, F. (2021). Study of Full-Body Virtual Embodiment Using Noninvasive Brain Stimulation and Imaging. *International Journal of Human-Computer Interaction, 37*(12), 1116–1129.

Steffen, J. H., et al. (2019). Framework of Affordances for Virtual Reality and Augmented Reality. *Journal of Management Information Systems, 36*(3), 683–729.

Terry, Q., & Keeney, S. (2022). *The Metaverse Handbook: Innovating for the Internet's Next Tectonic Shift.* Wiley.

Tsai, S.-P. (2020). Augmented Reality Enhancing Place Satisfaction for Heritage Tourism Marketing. *Current Issues in Tourism, 23*(9), 1078–1083.

Tulloch, R., & Johnson, C. (2022). Games and Data Capture Culture: Play in the Era of Accelerated Neoliberalism. *Media, Culture & Society, 44*(5), 922–934.

Tussyadiah, I. P., et al. (2018). Virtual Reality, Presence, and Attitude Change: Empirical Evidence from Tourism. *Tourism Management, 66*, 140–154.

van Rijmenam, M. (2022). *Step into the Metaverse: How the Immersive Internet Will Unlock a Trillion-Dollar Social Economy.* Wiley.

Wang, Y., Wang, L., & Siau, K. L. (2024). Human-Centered Interaction in Virtual Worlds: A New Era of Generative Artificial Intelligence and Metaverse. *International Journal of Human-Computer Interaction.* https://doi.org/10.1080/10447318.2024.2316376

Wong, M. (2020). Hidden Youth? A New Perspective on the Sociality of Young People 'Withdrawn' in the Bedroom in a Digital Age. *New Media & Society, 22*(7), 1227–1244.

CHAPTER 6

Where Are We At? Reconsidering (Un)Reality

Abstract While both the progressivist and the nostalgic responses to technological change have been shown to represent coping strategies that, in their own ways, merely reaffirm 'business as usual' (anthropocentrism as the default model), this chapter seeks alternative strategies to deal with the matter. Instead of opting for either of the extremes, a posthumanism-inspired solution focusing on interactivity and co-constitution of humans and their environment (including technology) is proposed. Such a horizontal model is claimed to avoid binary and hierarchical thinking characteristic of the mainstream anthropocentric thought. In this way, the aforementioned perspective is postulated as not only offering a viable explanatory apparatus but also being ethically preferable to the more established modes of thought.

Keywords Posthumanism • Co-constitution • Interaction • Flat terrain

While the progressivist and nostalgic points of view offer a seemingly stark choice, there is a need to go beyond such dualistic thinking by emphasising that choice is not only between the two coping strategies. Indeed, a more nuanced take on human-digital interaction would likely go behind

I. Kalpokas, *Technological Governance and Escapism in Times of Accelerated Change*, Information Technology and Global Governance, https://doi.org/10.1007/978-3-031-60890-2_6

the imaginary of '[s]parkling, spotless and new' data-infused capacities that 'promise a future that is scientifically perfectible, assuring a pot of gold at the end of a digitally coded rainbow' (Elish, & boyd, 2018: 58) or, on the contrary, nothing but misery and subjection or even destruction of humanity. Similarly, for Hepp (2020: 119), the mythology surrounding big data is seen to create the unrealistic expectation of 'rejuvenated access to the social world' and an unprecedentedly deep understanding of the processes involved without any need for engagement with theory or other pre-existing assumptions, thus ultimately driving substantial investment in the collection, processing, and further use of data (Hepp, 2020: 119)—a development which could be taken as representing a turn for either better or worse. Against the all-or-nothing approaches, it has to be emphasised that, in many ways, humans and algorithms are engaged in relationships of co-constitution, whereby not only algorithms shape humans and their worlds but also humans shape algorithms that learn from actual behaviours (Bucher, 2018: 94–95; see also Velkova & Kaun, 2019: 2). The focus thereby shifts towards interactive iterations and interactions take place whereby 'users neither passively submit to the guidance/control of algorithms, nor remain completely unaffected' but, instead, any effects and changes 'are co-constructed through ongoing "negotiations" between the two' (Karakayli et al., 2018: 6). Instead of focusing on the agency of either one or the other, one should instead look into collaborative agencies that are jointly achieved through interactions in always-ongoing acts of mutual composition.

6.1 Will the Real *Anthropos* Please Stand Up?

As already set out in this book, claims around human exceptionalism do appear problematic—indeed, unsustainable, if they ever were (Braidotti, 2019a: 35). After all, it transpires, in today's world at least, that the subject 'cannot be understood on the basis of identity, causally determined from the living being' (Goriunova, 2019: 139). There seems to be an ever-deepening connectedness that encompasses not just media infrastructures but also 'the layering of connected media practices in which individuals, collectivities and organisms are now routinely engaged' (Hepp, 2020: 85). Dewandre (2020: 13) also stresses the increasingly decentred and relational nature of the self that renders individuals vulnerable to malfunctioning of the system and the misdeeds of others; in this way, humans are left with no capacity to (re)gain the idealised rational independent functioning capacity—if that was ever possible. It is not untypical to claim that '[o]ne cannot think of humanity as distinct from technology anymore' (Stiegler, 2021: 203). This statement, however, only makes sense if one assumes there having been a time of idealised independence, largely corresponding with the emphasis on Cartesian mind-body dualism discussed in the previous chapters. Here, again, the independently active rational human mind is abstracted from the geographically and technologically placed body. The imaginary thereby created is one of a mind that is separate not only from its embodiment but also from any outside influence—until, that is, the advent of big data and algorithms. In this way, the narrative is once again one of, in a way, paradise lost—in other words, nostalgia. While not always openly shaped as nostalgic, such accounts nevertheless focus on the loss or undermining of the Cartesian cogito, particularly within the current technological context (see e.g. Aydin, 2021: 101).

As noted previously in the book, there is a strong aura of unavoidability in accounts of technological change—as if there was only one, predetermined, trajectory into the future whereby no outside of the digital remains. As Susskind (2018: 118) puts it, every aspect of daily life is becoming subjected to technology's power to the extent that '[t]he person who tries to opt out of code's empire will be like the man Hume describes as trying to opt out of the state itself; carried aboard a ship while asleep. His only means of escape is to jump overboard'. Similarly, for Greenfield (2018: 81-82), it is those left behind, either due to existing inequalities or due to conscious objection that will bear the brunt of the cost by either being decoupled from the increasingly digital-first world or by way of being rendered inferior to their increasingly technologically augmented counterparts; in either case, their status would be undermined. Power in the society of the near future will rest with data (if that is not the case already). In this sense, it can be expected that with datafication becoming the norm, it will be 'control of the *means of scrutiny*' that will characterise peak control (Susskind, 2018: 124). Here, of course, a clear shift is visible: from Marx's emphasis on the ownership of the means of production to ownership of the means of scrutiny. Indeed, regardless of one's position on the nostalgia—progressivism continuum, it must be admitted that control of the data flows and analytics will be a crucial source of power in a future society with issues of accountability and democratic control looming particularly high. Simultaneously, though, issues pertaining to the subject who is supposed to carry out such control—effectively, the *anthropos* of anthropocentrism—cannot be ignored either.

Crucially, the 'human' of humanism (and, hence, also the *anthropos* of anthropocentrism) itself is not a straightforward and self-evident entity, with the meaning and content of the ideal of the human being contingent upon historical, cultural, ideological etc. shifts (Braidotti, 2013: 1; Herbrechter, 2013: 38; Coeckelbergh, 2020: 42). It must, therefore, be stressed that '[a]ppeals to the "human" are always discriminatory: they create structural distinctions and inequalities among different categories of humans, let alone between humans and non-humans' (Braidotti, 2019a: 35). And while some aim to defend humanity, stressing how any such critique only applies to those members of humanity who discriminate non-humans, just like 'the label "sexist" or "racist" does not apply to all males or all whites' (Kopnina et al., 2018: 122), these attempts evade the historical and cultural embeddedness of anthropocentrism (or sexism or racism—as well as relations between those discriminatory practices). Instead, a more critical reading of the limits of 'the human' implicated in anthropocentrism is merited in order to further clarify the loci of discriminatory practices (including with regard to truth and meaning) not just associated with but also, more importantly, foregrounding 'the human'.

Simultaneously, it must be remembered that while it is true that humanity had been used as the norm and measure of all things, one must think more carefully about what that exalted notion of humanity has historically entailed (and what in entails today). Even if one considers the 'human' (*anthropos*) as the core figure of Western modernity and rationalist Enlightenment heritage, one must keep in mind that this human is not (and has never been) *any* human but, instead, merely a privileged segment of humanity. For the most part, it was primarily the affluent white Western male who could afford (in terms of both resources and power) the imaginary capacity for Cartesian disembodied detachment at the expense of *his* less-privileged, non-privileged, and under-privileged counterparts, including both other humans and the entirety of the non-human world (Grear, 2015; Dekeyser, 2023). Hence, as stressed by De Lucia (2017: 189), anthropocentrism concerns, first and foremost, 'a particular subset of humanity that historically has colonized, exploited and plundered both other humans and the non-human world'. The artificiality of the Cartesian *man* (inclusive of the sexist connotations) means that any account of reality is condemned to be either nostalgic or utopian because they unavoidably revolve around an absent centre.

It is also worth noting that a homogenising drive has been a constant feature even of humanity's internal view of itself. While the racialised and gendered premises of the 'human' benchmark are evident, a notable contribution is also made by critical disability studies, particularly through their refusal to treat disability as pregiven and objective, but focusing, instead, on people as 'disabled by interrelations between ability and disability' (Flynn, 2021: 636). No less importantly, *despite* such contingency, processes of exclusion were never less strict—instead, a characteristic feature of maintenance of 'the human' has long been '[t]he pathologisation, criminalisation, dehumanisation – and by extension – exclusion from citizenship of those who statistically deviate from the norm is culturally bound and historically specific' (Kiwan, 2022: 533). This, however, is not only relevant to the argument put forward in this book in an abstract sense (as yet another example of exploitative and discriminatory practices inherent in 'the human') but also more directly due to historical changes of the norms of humanness and (dis)ability through technological augmentation and a prominent drive towards 'techno-ablism' as a means to 'fix' disability (Kiwan, 2022: 535–536). In other words, (in)capacities tend to be defined vis-à-vis the dominant technological frameworks and the qualities needed to productively function within the latter (for example, algorithms today reinforcing the 'normal' through hegemonic representations in data), while technology is simultaneously seen as a means of overcoming that same frailty and 'deficiency' of the embodied human.

The preceding paragraphs may initially seem as contradictory: on the one hand, there is no outside of responsibility for discriminatory and exploitative practices and, secondly, the 'humanity' that has performed and benefitted from actual exploitation has only been a privileged majority. However, both sides of the coin are best seen as contributing to the whole. On the one hand, the particularity of 'the human' serves to show the futility of the pretence of the allegedly universal human figure who serves as a measure of all things and of humans as allegedly having superior ontological, epistemological, and ethical value. Instead, the preceding is unmasked (and reflected through digital technologies) as an always already skewed representation of reality. Simultaneously, though, this does not render the (previously or currently) privileged groups solely responsible for the uprooting of such practices. Instead, creation of more egalitarian approaches should be seen as everybody's shared responsibility, although the issue of means- and track record-based contributions might seem to be a reasonable one.

Nevertheless, 'the human' has typically been used as a boundary marker between those who had self-arrogated the status as humans and those deemed less- or even un-worthy. When coupled with the nostalgic sentiment, such thinking implies the return of (parts of) society into some kind of primitive, almost child-like state inferior to the imaginary standard whereby humans are effectively nursed by AI. Crucially, references to such ascriptions of primitiveness and irrationality and their contrasting with objectivity and reason have been part and parcel to the Enlightenment legacy underpinning discrimination and colonialism. Naturally, tackling the humanist subject that serves as a grounding for violence and exclusion is a priority. And yet, one must also agree with Geerts (2022) that it would be naïve to expect that, given the history of violence and exploitation, a flat ontology untainted by power relations and unhindered by historical injustices would suddenly be possible and complete equality would immediately take shape. Nevertheless, posthumanism at least offers a normative ideal of interactive co-occurrence that ought to be positioned as normatively superior to any hierarchical alternatives or half-way measures.

6.2 How to Draw a Flat Terrain

On the one hand, it is becoming clear that 'the very cogito that Descartes rendered independent and pure is shaped by the objects and artifacts that surround it', with technological artefacts and infrastructures being 'a major force that not only is increasingly shaping our world, but also is shaping our mind and self' (Aydin, 2021: 101). In this way, technologies are seen as becoming 'extimate', that is, both deeply engrained within the human self and enacting shaping from outside (Aydin, 2021: 243). On the other hand, though, it is crucial not to give in to one-sided interpretations of such shaping, particularly since the outside is equally extimate as well. Likewise, while traditional, essentialist, definitions of human agency portray the latter as a zero-sum game, particularly when it comes to human-technology interaction, they ignore the ways in which shared and jointly situated agency can be produce through the interactions between users and technological artefacts, negotiating mutual affordances as they go (Pentzhold & Bischof, 2019: 7). In this, way, then, '*the effects of algorithms can be reshaped through user practices*' (Karakayli et al., 2018: 6), pointing towards a more interactive and egalitarian mode of interaction,

which is also less prone to either of the two extremes-focused coping strategies.

In order to arrive at an ethically justifiable interpretation of human-machine interaction that does not fall into either the nostalgic or the progressivist essentialisms, it is important to go 'beyond the competition narrative about humans and machines' (Coeckelbergh, 2020: 42). However, contra Coeckelbergh, this should not be taken as somehow inextricably tied with deliberate border crossing, particularly because the very idea of there being such borders is one that should be overcome as a vestige of the misguided dualism inherent in Western modernity's misconstruction of the world. In fact, such an illusory boundary between the individual and the world, if left unresolved, 'leaves us trapped in the illusion of isolation' (Oliver, 2020: 136). Crucially, then, instead of an independent autonomous self that is capable of rational thought and introspection in line with the Cartesian cogito, one should consider the conscious self as simply 'the sequence of *outputs* of a cycle of thought, locking onto, and imposing meaning on, aspects of the sensory world' (Charter, 2019: 9). Likewise, nonhuman entities need to be taken seriously, specifically highlighting, in the context of this book, 'device parameters and affordances, analytical algorithms, data infrastructure, and data itself, as well as the processes and practices around them' (Sandvik, 2020: 2017), although by no means predetermined to one-sidedly take away human agency.

Crucially, instead of well-defined entities that need to transcend identifiable boundaries, humans 'are always already distributed phenomena, interembodied with other humans and with nonhumans, multiple and open to the world' (Lupton, 2020: 14). In such contexts, emphasis is on '*co-constitution* of home, gender, familial relationships and digital technologies' (Watson et al., 2021: 1210). Hence, an entity can only be known and understood through referring to its history and patterns of interaction within the specific environment within which it is present, including with other adjacent and otherwise related entities—which can only mean that any agent can only be accounted for and comprehended 'in a specific sociotemporal situation' (Heras-Escribano, 2019: 2; see also Charter, 2019: 10). Simultaneously, though, it must be kept in mind that the process is not one-way either: both humans *and* their counterparts, such as algorithms, come to act only in and through (inter)relationships (Amoore, 2020: 8). Likewise, algorithms also should not be seen as standalone entities but as enmeshed and interconnected—after all, their functioning is contingent upon the historical and cultural context in which the data on which they are trained had been collected and also the context within which the datasets used for the training has been compiled from the pools of available data (Bechmann, 2019: 82–83). Indeed, it is the relationality and affective capacity that should be seen as the defining qualities of any entities, human ones included (Braidotti, 2019b: 45; see also, albeit from a different perspective, Oliver, 2020: 91). With the latter in mind, what needs to be overcome in order to develop a more ethical approach towards the self and the world is culturally embedded egocentrism, that is, 'the illusionary belief that we are a coherent self at the centre of a universe'; nevertheless, it is imperative not just that the self is decentred but also that it is deconstructed, recast as 'fully enmeshed within' the surrounding world (Oliver, 2020: 241).

Indeed, any distinctions, even if at all theoretically possible, become indeterminate in the context of 'human-data assemblages' (Lupton, 2020: 17). The concept of assemblage is particularly useful in this context, referring to 'gatherings of human and nonhuman agents, or a collection of things' without them assuming any permanent characteristics but being, instead, 'fluid and mobile, constantly coming together and coming apart' (Lupton et al., 2022: 365). As Ash (2015: 85) claims, 'to be human is to be shaped by and immersed in technical affects', with the latter understood as 'a vital force generated with and through assemblages, or gatherings, of humans and nonhumans' (Lupton et al., 2022: 365). What transpires to be strictly impossible, therefore, is a detached representation of what 'really is', as if from some objective vantage point—instead, any momentary iteration of the 'is' is only achieved through ongoing practices and interactions (Mauthner, 2019: 671).

Attention here has to be paid to the notion of affordances, typically understood as 'the mutual constitution between people and environments' (Davis, 2020: 6)—although, if one takes the posthumanist predicament seriously, the singling out of 'people' should be seen as overly restrictive. Instead, it would be more reasonable to stress the mutual and simultaneous provision and diminishing of capacities within the interaction between any agents as well as between such agents and their environments. Likewise, identities and (self-)representations become closely tied to and only possible through app, algorithm, and platform affordances (Cirucci, 2023). Focus on affordances is an ethically justifiable perspective because it helps strike the balance between agency and materiality (Davis, 2020: 7) as well as digitality. Crucially, it must be kept in mind that affordances are never stable or predetermined but, instead, achieved through a process of interaction, including, but by no means necessarily, between humans and technological artefacts (Pentzhold & Bischof, 2019: 5–6). The preceding also has a clear effect on agency, essentialist and binary definitions of which 'fail to account for the association of people and robots in the collective accomplishment of affordances' through common partaking in actions (Pentzhold & Bischof, 2019: 7). It must, therefore, be acknowledged that 'meaning and matter, the social and the technological, are inseparable and they do not have inherently determinable boundaries and properties' but are, instead, 'constituted as relational effects performed in a texture of situated practices' (Gherardi, 2012: 40). Consequently, everyday life is not a measurable trajectory but, instead, 'an ongoing composition in which humans and non-humans participate' (Neyland, 2019: 11). Once again, a flat ontology of relationality appears as an alternative to the nostalgic or progressivist essentialisms.

Crucially, then, it transpires that 'humans and machines have the potential to be theoretically and operationally equivalent agents' as both are identically engaged via 'agency, interactivity, and influence' (Banks & de Graaf, 2020: 29). In this way, a pathway is opened up for a de-centred world the flourishing of which is premised on 'co-becomings with the multifarious creatures in more-than-human worlds', which also means a horizontal and egalitarian sharing of responsibilities (Lupton et al., 2022: 365). This should not be confused with paternalistic understandings of responsibility whereby one becomes responsible for someone or something as an act of benevolence, from a position of power. Instead, this is a responsibility based on shared interdependence and a shared ability to respond, even if such responses may come in varying ways, shapes, and forms. What the above should not be taken to account for is some kind of cyborgisation of the human, despite some nominally posthumanist authors suggesting so (see e.g. Nath & Manna, 2023). What cyborgisation entails is merely a merger of human and machine essentialisms to produce a novel condition that had never been before. What one should focus on, instead, is that human independence and isolation had always been a misconstruction and not just something currently being brought about by technological augmentation (the latter also being easily susceptible to progressivist overtones).

Even the idea of personality and a sense of self cannot be taken for granted or as stable, but instead works in a relational way. Here, the point in question are the ways in which human selves are affected by artificial personalities while the aforementioned personalities are co-constituted in the interactions between pre-set features of such virtual agents and human user projections and behaviours. Although bumping into human-like robots in the street or any other physical environment might well remain a rarity in the short term, social media platforms and virtual environments are the domains in which not only humans and human-like robots mix—it is also a domain where robots (as virtual influencers) are followed by humans (Mosley, 2020; see also Haikel-Elsabeh, 2023).

Certainly, it is possible to trace a certain progression of (un)reality, whereby the distinction between reality and image has first been erased by marketing, then the line between real and manufactured identity was erased by social media; once those two converge, the logical next step would be the creation of virtual personalities (Turner-Williams, 2020). While people may have learned to doubt the authenticity of other humans online, and influencers in particular, there usually at least exists a reason to believe that there is at least some substance behind the content. Simultaneously, though, trust in virtual influencers transpires to be related to them being 'authentically fake' (Lou et al., 2023; see also Allal-Chérif et al., 2024), which might be seen as a welcome counterweight to the strategic fakery of online human behaviours.

Nevertheless, the growing popularity of virtual influencers seems to indicate that people do not particularly care about their lack of conventional reality, and are likely to care less and less. It is all in the eyes of the beholder, and as long as the audience identifies with what they see and hear, the question of reality becomes suspended. That might explain why even virtual influencers that are decidedly cartoonish and unrealistic manage to craft their niche, including marketing very real goods (Sokolov, 2019). Also, while it is technically correct to point out that a virtual influencer is, after all, 'nothing more than a coded puppet at the hand of the people and brands who control it' (Haile, 2020), it is the invested meaning that really counts. And, as a matter that is further down the line, but nevertheless relevant, a question must be asked as to what happens when some degree of AI-based autonomy kicks in. This may, in the foreseeable future at least, not be anything approximating full autonomy, but rather (semi-)automated appearance and behaviour rendering, based on the available audience data and geared simply towards maximising engagement. Nevertheless, the immediacy (and, in a way, authenticity) of audience engagement is going to be transformed even further, pushing the boundaries of identification, emotional investment, and (perceived) reality.

On the one hand, it transpires that such virtual persons only take the escapist tendencies to another level by explicitly rejecting reality as such (Mirowska & Arsenyan, 2023). In fact, both progressivist and nostalgic interpretations of such escapism can be easily conceived. On the progressivist side, the emphasis would be on the automation of human sociality and the benefits inherent therein, from enhanced emotional care to enhanced sales through virtual influencers. Meanwhile, on the nostalgic side, the loss of 'humanness' in yet another crucial area of life would be lamented and the advent of a completely technology-engulfed and technology-driven life would be criticised. Nevertheless, engagement with virtual personas can also be seen in a completely different way—as a particularly pronounced (because virtually embodied) example of human-digital interactivity, one that takes the latter principle from an abstract to a very concrete level. What we have here is a matter of continuous exchange whereby human selves are affected, and sometimes even transform, though interactions with digital entities while those same entities are simultaneously constantly adjusting to human-derived data, thereby building a bond of mutual co-constitution on a flat terrain (Fig. 6.1).

What hereby emerges is a continuous loop stretched across a flat terrain. Both aspects are important here: as a loop, it resists dichotomies while by resting on a flat terrain it cannot succumb to hierarchies of any

Fig. 6.1 Human-digital exchange

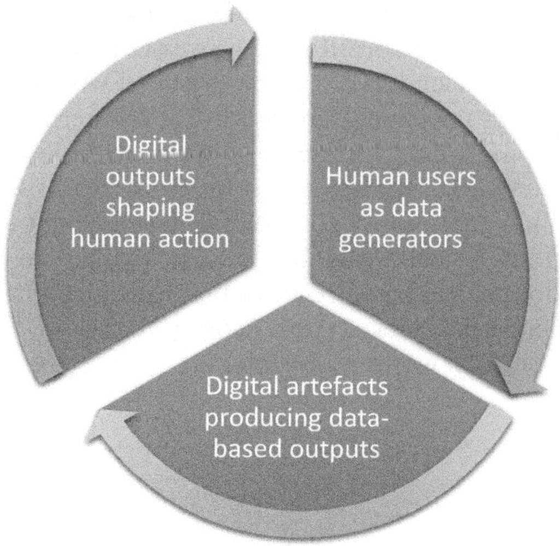

sorts. In this interrelationship, humans are open to the absence of priority and autonomy (which, to reiterate were always imaginary anyway) but, instead, act as generators and sources of data. That, however, *does* imply some degree of activity and determining power because it is from such data that digital outputs learn and adjust themselves in order to better fit the needs of the human users. In a way, therefore, this part of the loop might even appear as if it was inching closer towards the anthropocentric ideal. Nevertheless, digital artefacts are not *only* learning and adjusting, but also shape human actions and behaviours in return by structuring the environment and conditions for the human self to unfold, which is where the human openness to being partly subjected to outside really kicks in. Consequently, both humans and digital artefacts simultaneously are active and passive, determining and determined.

A question might be raised, though, as to why there is an intermediary shaping step between machines and humans and none between humans and machines. The reason for that is technological efficiency. While the datafication of human behaviours, data analytics, and the shaping of digital artefacts happens in an automated fashion, the same (still) does not apply to the return phase of this loop. Humans are still shaped indirectly: through interactions with virtual agents, through the digital architectures of interfaces, through curation of the information environment etc. Of course, there are steps being taken to automate this process as well, but until the mainstreamisation of, for example, brain-computer interfaces and similar technologies, an intermediary step remains necessary and the mirroring of co-constitution processes remains incomplete.

In the end, it can be claimed that while reliance on a dualist view of the world, deeply entrenched in the Western tradition, remains an attractive coping strategy in times of rapid technological change, an alternative and alternative, non-binary strategy is also possible. Not only it is one that allows for a more constructive engagement with the technologised conditions of daily life, but also it enables an avoidance of a hierarchical, dominance-based relationship, which has characterised Western anthropocentrism. Hence, ethical coexistence with both the natural and the technological world *is* possible.

References

Allal-Chérif, O., Puertas, R., & Carracedo, P. (2024). Intelligent Influencer Marketing: How AI-Powered Virtual Influencers Outperform Human Influencers. *Technological Forecasting and Social Change, 200*, 1–14.

Amoore, L. (2020). *Cloud Ethics: Algorithms and the Attributes of Ourselves and Others*. Duke University Press.

Ash, J. (2015). Technology and Affect: Towards a Theory of Inorganically Organised Objects. *Emotion, Space and Society, 14*, 84–90.

Aydin, C. (2021). *Extimate Technology: Self-Formation in a Technological World*. Routledge.

Banks, J., & de Graaf, M. M. A. (2020). Toward an Agent-Agnostic Transmission Model: Synthesizing Anthropocentric and Technocentric Paradigms in Communication. *Human-Machine Communication, 1*, 19–36.

Bechmann, A. (2019). Data as Humans: Representation, Accountability, and Equality in Big Data. In R. F. Jorgensen (Ed.), *Human Rights in the Age of Platforms*. The MIT Press.

Braidotti, R. (2013). *The Posthuman*. Polity Press.

Braidotti, R. (2019a). A Theoretical Framework for the Critical Posthumanities. *Theory, Culture & Society, 36*(6), 31–61.

Braidotti, R. (2019b). *Posthuman Knowledge*. Polity.

Bucher, T. (2018). *If... Then: Algorithmic Power and Politics*. Oxford University Press.

Charter, N. (2019). *The Mind Is Flat: The Illusion of Mental Depth and the Improvised Mind*. Yale University Press.

Cirucci, A. M. (2023). Zoom Affordances and Identity: A Case Study. *Social Media + Society*. https://doi.org/10.1177/20563051221146176

Coeckelbergh, M. (2020). *AI Ethics*. The MIT Press.

Davis, J. L. (2020). *How Artifacts Afford: The Power and Politics of Everyday Things*. The MIT Press.

De Lucia, V. (2017). Beyond Anthropocentrism and Ecocentrism: A Biopolitical Reading of Environmental Law. *Journal of Human Rights and the Environment, 8*(2), 181–202.

Dekeyser, T. (2023). Rethinking Posthumanist Subjectivity: Technology as Ontological Murder in European Colonialism. *Theory, Culture & Society*. https://doi.org/10.1177/02632764231178482

Dewandre, N. (2020). Big Data: From Modern Fears to Enlightened and Vigilant Embrace of New Beginnings. *Big Data & Society*. https://doi.org/10.1177/2053951720936708

Elish, M. C., & boyd, d. (2018). Situating Methods in the Magic of Big Data and AI. *Communication Monographs, 85*(1), 57–80.

Flynn, S. (2021). Corporeality and Critical Disability Studies: Toward an Informed Epistemology of Embodiment. *Disability & Society, 36*(4), 636–655.

Geerts, E. (2022). Navigating (Post-)Anthropocentric Times of Crisis: A Critical Cartography of Hope. *CounterText, 8*(3), 385–412.

Gherardi, S. (2012). *How to Conduct a Practice-Based Study: Problems and Methods*. Edward Elgar.

Goriunova, O. (2019). The Digital Subject: People as Data Persons. *Theory, Culture & Society, 36*(6), 125–145.

Grear, A. (2015). Deconstructing *Anthropos*: A Critical Legal Reflection on 'Anthropocentric' Law and Anthropocene 'Humanity'. *Law and Critique, 26*, 225–249.

Greenfield, A. (2018). *Radical Technologies: The Design of Everyday Life*. Verso.

Haikel-Elsabeh, M. (2023). Virtual Influencers Versus Real Influencers Advertising in the Metaverse, Understanding the Perceptions, and Interactions with Users. *Journal of Current Issues & Research in Advertising, 44*(3), 252–273.

Haile, J. (2020, September 18). What Are the Ethical Implications of CGI Influencers Promoting Brands? *Talking Influence*. https://talkinginfluence.com/2020/09/18/ethical-implications-cgi-influencers/

Hepp, A. (2020). *Deep Mediatization*. Routledge.

Heras-Escribano, M. (2019). *The Philosophy of Affordances*. Palgrave Macmillan.

Herbrechter, S. (2013). *Posthumanism: A Critical Analysis*. Bloomsbury Academic.

Karakayli, N., Kostem, B., & Galip, I. (2018). Recommendation Systems as Technologies of the Self: Algorithmic Control and the Formation of Music Taste. *Theory, Culture & Society, 35*(2), 3–24.

Kiwan, D. (2022). Dis/abled Decolonial Human and Citizen Futures. *Citizenship Studies, 26*(4), 530–538.

Kopnina, H., et al. (2018). Anthropocentrism: More than Just a Misunderstood Problem. *Journal of Agricultural and Environmental Ethics, 31*, 109–127.

Lou, C., et al. (2023). Authentically Fake? How Consumers Respond to the Influence of Virtual Influencers. *Journal of Advertising, 52*(4), 540–557.

Lupton, D. (2020). *Data Selves*. Polity.

Lupton, D., Clark, M., & Southerton, C. (2022). Digitized and Datafied Embodiment: A More-than-Human Approach. In S. Herbrechter et al. (Eds.), *Palgrave Handbook of Critical Posthumanism* (pp. 361–383). Palgrave Macmillan.

Mauthner, N. S. (2019). Toward a Posthumanist Ethics of Qualitative Research in a Big Data Era. *American Behavioral Scientist, 63*(6), 669–698.

Mirowska, A., & Arsenyan, J. (2023). Sweet Escape: The Role of Empathy in Social Media Engagement with Human Versus Virtual Influencers. *International Journal of Human-Computer Studies, 174*, 1–18.

Mosley, M. (2020, July 14). Virtual Influencers: What They Are and How Do They Work? *Influencer Matchmaker*. https://influencermatchmaker.co.uk/blog/virtual-influencers-what-are-they-how-do-they-work.

Nath, R., & Manna, R. (2023). From Posthumanism to Ethics of Artificial Intelligence. *AI & Society, 38*, 185–196.

Neyland, D. (2019). *The Everyday Life of an Algorithm*. Palgrave Macmillan.

Oliver, T. (2020). *The Self Delusion: The Surprising Science of How We Are Connected and why that Matters*. Weidenfeld and Nicolson.

Pentzhold, C., & Bischof, A. (2019). Making Affordances Real: Socio-Material Prefiguration, Performed Agency, and Coordinated Activities in Human-Robot Communication. *Social Media + Society*. https://doi.org/10.1177/2056305119865472

Sandvik, K. B. (2020). Wearables for Something Good: Aid, Dataveillance and the Production of Children's Digital Bodies. *Information, Communication & Society*. https://doi.org/10.1080/1369118X.2020.1753797

Sokolov, M. (2019, December 5). Virtual Influencer Trends: An Overview of the Industry. *The Drum*. https://www.thedrum.com/opinion/2019/12/05/virtual-influencer-trends-overview-the-industry.

Stiegler, C. (2021). *The 360° Gaze: Immersions in Media, Society, and Culture*. The MIT Press.

Susskind, J. (2018). *Future Politics: Living Together in a World Transformed by Tech*. Oxford University Press.

Turner-Williams, J. (2020, May 28). Are Virtual Influencers More Valuable to Brands than Humans? *Super Maker*. https://supermaker.com/articles/virtual-influencers-changing-social-media

Velkova, J., & Kaun, A. (2019). Algorithmic Resistance: Media Practices and the Politics of Repair. *Information, Communication & Society*. https://doi.org/10.1080/1369118X.2019.1657162

Watson, A., Lupton, D., & Michael, M. (2021). The COVID Digital Home Assemblage: Transforming the Home into a Work Space During the Crisis. *Convergence: The International Journal of Research into New Media Technologies, 27*(5), 1207–1221.

CHAPTER 7

Conclusions

Abstract While the speed and scale of technological transformations necessitates explanatory efforts, as this final chapter concludes, such efforts typically come by way of occluding key tensions within conceptions of the role of the human and their relationship with the environment. In fact, such occlusion is only possible if not only the present but also the past and, correspondingly, the future are lost track of, thereby opening up the way for imaginaries informed by predominant expectations and fears. As such narratives are seen as disempowering, the interactive alternative proposed in this book gains additional value.

Keywords de-politicisation • time perspective • disempowerment

The rise in prominence and growing adoption of AI and related technologies has been one of the defining features of contemporary societies. For the most part, such technologies are already neatly integrated into daily lives, thereby creating a sense of smoothness, of progress already underway and likely to slide, in the same smooth manner, into foreseeable future. However, no process can function entirely smoothly and without friction, either due to its own internal brittleness that occasionally comes

I. Kalpokas, *Technological Governance and Escapism in Times of Accelerated Change*, Information Technology and Global Governance, https://doi.org/10.1007/978-3-031-60890-2_7

to the fore or as a result of the environment in which it unfolds. Once such frictions become prominent, however, the progress argument tends to be undermined and the value of, in this case, technological change becomes a matter of heated debate. As a result, then, any snapshot of the present is caught in a tension between smoothness and friction. While, as has been argued in this book, this tension is not necessarily the most productive outcome, it is, nevertheless, an intuitive one.

In light of the above tension, this book has identified two broad groups of approaches towards technological development: a nostalgic one, focusing on the alleged loss of human exclusivity, autonomy, capacity for rational though, and related qualities and a progressivist one, stressing the abundance of wealth and new possibilities arising from emerging technologies and the improvements they would bring not only to the economic and social domains but even to the human as such. Both, nevertheless, are seen primarily as strategies for coping with the loss of the anthropocentric ideal, characteristic of the Western tradition. While, it transpires, humans had never lived up to that benchmark anyway (and it is in itself far from morally justifiable), its loss is still acutely felt—perhaps precisely *because* it is the loss of an absent centre.

At the heart of the drive towards either the nostalgic or the progressivist visions of technology is a challenge that the joint thrust of platformisation, datafication, and AI poses to long-standing cherished assumptions about human primacy, exceptionality, and autonomy. Such assumptions have been shown to stem from long-standing patterns of anthropocentric thought that is deeply engrained within the Western intellectual tradition. In order to counter this status quo, posthumanist critique is used as a point of departure. The nostalgia-progressivism tension has then been traced across the dominant accounts of post-truth, which are shown to largely succumb to the nostalgic interpretation of a technological degrading of the human, interpretations of technological innovations more broadly, which are shown to be split between the two poles, and visions for the metaverse and other forms of technological escape that quintessentially embody the progressivist version of anthropocentrism. In the end, effort has been put to develop an alternative to such binary thinking.

With the above in mind, it is unsurprising that Sartori and Bocca (2023: 455) call for 'a sociotechnical perspective in the study of technology'—one that 'balances dominant deterministic approaches'. However, while the *intention* here is generally commendable, the outlook is less so: after all, balancing is only possible between alternatives that stand on a solid ground, and not on a groundless terrain of anthropocentric assumptions. Likewise, balancing typically involves a compromise or even a synthesis while the outlook proposed in this book is intended to serve as an *alternative* to the entire progressivism-nostalgia dichotomy. Therefore, this book has strived to widen the scope of possible approaches and provide a an ethically justifiable outlook that avoids either a furthering of or a nostalgia for anthropocentric patterns of domination and, consequently, resists the temptation to merely gloss over the differences of such equally unsustainable points of view.

Cheney-Lippold (2024) stresses the problem of the past having been forgotten and the present being demoted. When the past has been forgotten, everything seems new and unprecedented; and when the present is demoted, salvation is sought in the future. Moreover, the inability to see inspiration in the past and possibility of meaningful action in the present removes any impetus (or even possibility) for any action here and now and, therefore, denies the ability to shape directions for the future. Instead, one is likely to succumb to technological determinism, since whatever comes in the future not only *will* come but will also be (it seems) more impactful and meaningful than anything that is, for better or for worse. On the other hand, denigration of the present and forgetfulness about the past can lead to the opposite focus: yearning for an *imagined* or, at least, whitewashed, sanitised, and idealised past where the ills of the present (and of the projected future) had not yet taken shape (see, albeit with a different agenda, Tanner, 2020). Both outcomes, then give rise to their own escapist coping strategies.

For others, technology acts as a 'de-politicization machine' (Elwood, 2021: 216–217), implying not only unavoidability of the digital future but also digital solutions in the present being understood as so 'smart' they cannot be contested. Crucially, one must also keep in mind that technological development, including the digital architecture of the code and the physical computational and communicative infrastructure, are underpinned by, and cannot be imagined without, the industrial scale and corporate interests of the companies active in the field (see e.g. van der Vlist et al., 2024). From the commercial standpoint, then, the sense of technological development and ever growing adoption as taken for granted has obvious benefits. Such aura of unavoidability, it has been argued in this book, contributes to the perceived urgency of taking sides on the nostalgia-progressivism dichotomy.

It must be noted that both the nostalgic and the progressivist escapisms are ultimately disempowering—as all escapisms are—because they distract from the present and from possible struggles for a different future. Therefore, this book has sought an alternative solution that would go beyond dichotomic thinking. An interactive model of co-constitution is thus developed in which humans and digital artefacts (although, going beyond the ambit of this book, natural artefacts as well) are all locked in an interactive loop stretched across a flat terrain. As a model that is neither hierarchical nor giving a special role or privilege to either an anthropocentric interaction or its constitutive parts, the proposed framework is claimed to offer an ethically justifiable alternative to the two escapist coping strategies.

No less importantly, fixation on one specific socio-technical imaginary, or even on a particular dichotomy, leads to 'discursive closure', namely, a situation when 'certain types of communication practices shut down or close off options for thinking otherwise' (Markham, 2021: 392). In a similar manner, for Hoff (2023: 11), the governance of AI (but also, by extension, of other digital technologies as well) pertains to not only formal regulations but also 'the management of affects and emotions concerning the future of AI'. In this sense, a notable thrust of this book has been opening up a space for thinking otherwise and enabling a broader scope of affective and emotional entanglements. This has been achieved by way of building on broadly posthumanist critiques of anthropocentrism and focusing specifically on human entanglement with the technological environment and the interactivity of this relationship.

Refusal to be content with discursive closure is key because of the role sociotechnical imaginaries play in shaping individuals' attitudes and behaviours. As argued by Sörum and Fuentes (2023: 41), 'sociotechnical imaginaries frame how consumers interpret, know, feel, act and make sense' of technological processes and innovations while also shaping 'current and future views on and digital practices with digital devices'. Similarly, as noted by Sartori and Theodorou (2022: 6), the dominant narratives, by way of circulation and repetition, affect both the understanding and the reception of the technologies in question (for a similar argument, see also Vrabič Dežman, 2024). In this way, a closure that leaves certain imaginaries as default significantly restricts not just attitudes but also actual and potential practices and behaviours. Hence, looking beyond established binaries is also crucial in terms of encouraging attitudinal and behavioural change.

Again, merely thinking otherwise is insufficient. Here, a notable example could be the call to '*rehumanise* automation', by which the authors mean 'establishing the human as a critical and creative agent in sociotechnical transformations and human–machine relationships' (Pink et al., 2022: 3). As the wording makes clear, such a call for new imaginaries remains strictly within the anthropocentric frame of reference. For this reason, the call for rejecting anthropocentric patterns of domination and exploitation and untethering sociotechnical imaginaries from such ethically and empirically unsustainable assumptions is key: only if such an alternative course is taken, a more productive engagement with technology is possible.

Bibliography

Cheney-Lippold, J. (2024). The Silicon Future. *New Media & Society*. https://doi.org/10.1177/14614448241234864

Elwood, S. (2021). Digital Geography, Feminist Relationality, Black and Queer Code Studies: Thriving Otherwise. *Progress in Human Geography, 45*(2), 209–228.

Hoff, J.-L. (2023). Unavoidable Futures? How Governments Articulate Sociotechnical Imaginaries of AI and Healthcare Services. *Futures, 148*, 1–13.

Markham, A. (2021). The limits of the Imaginary: Challenges to Intervening in Future Speculations of Memory, Data, and Algorithms. *New Media Society, 23*, 382–405.

Pink, S., et al. (2022). Everyday Automation: Setting a Research Agenda. In S. Pink et al. (Eds.), *Everyday Automation: Experiencing and Anticipating Emerging Technologies* (pp. 1–19). Routledge.

Sartori, L., & Bocca, G. (2023). Minding the Gap(s): Public Perceptions of AI and Socio-Technical Imaginaries. *AI & SOCIETY, 38*, 443–458

Sartori, L., & Theodorou, A. (2022). A Sociotechnical Perspective for the Future of AI: Narratives, Inequalities, and Human Control. *Ethics and Information Technology, 24*, 1–11.

Sörum, N., & Fuentes, C. (2023). How Sociotechnical Imaginaries Shape Consumers' Experiences of and Responses to Commercial Data Collection Practices. *Consumption Markets & Culture, 26*(1), 24–46.

Tanner, G. (2020). *The Circle of the Snake: Nostalgia and Utopia in the Age of Big Tech*. Zero Books.

van der Vlist, F., Helmond, A., & Ferrari, F. (2024). Big AI: Cloud Infrastructure Dependence and the Industrialisation of Artificial Intelligence. *Big Data & Society*. https://doi.org/10.1177/20539517241232630

Vrabič Dežman, D. (2024). Promising the Future, Encoding the Past: AI Hype and Public Media Imagery. *AI & Ethics*. https://doi.org/10.1007/s43681-024-00474-x

Index

I. Kalpokas, *Technological Governance and Escapism in Times of Accelerated Change*, Information Technology and Global Governance, https://doi.org/10.1007/978-3-031-60890-2

SPRINGER NATURE

GPSR Compliance

The European Union's (EU) General Product Safety Regulation (GPSR) is a set of rules that requires consumer products to be safe and our obligations to ensure this.

If you have any concerns about our products, you can contact us on ProductSafety@springernature.com

In case Publisher is established outside the EU, the EU authorized representative is:

Springer Nature Customer Service Center GmbH
Europaplatz 3
69115 Heidelberg, Germany

The manufacturer's authorised representative in the EU is Springer Nature Customer Service Centre GmbH, Europaplatz 3, 69115 Heidelberg, Germany. If you have any concerns regarding our products, please contact ProductSafety@springernature.com

Printed and bound by CPI Group (UK) Ltd, Croydon, CR0 4YY
01/10/2026
02225879-0003